COGNITIVE ENGINEERING

Third Millennium Awareness Science

J.L. Mee

Cognitive Engineering
Third Millennium Awareness Science

ORIGINAL MANUSCRIPT

Limited Edition

Copyright © 2021 by J. L. Mee

All rights reserved, including the right to reproduce this book or portions thereof in any form. For information, contact the publisher.

Portions of this work originally appeared in an earlier book, *Future Life Design*, by J.L. Mee.

Mee, J.L.
   Cognitive Engineering – Third Millennium Awareness Science
   / by J.L. Mee

ISBN 978-1-892654-30-4

1. Spirituality   2. Consciousness   3. Neuroscience

Cover artwork by Andrew Ostrovsky

Printed in the United States of America

Future Life University Press

Notes to Readers of the Original Manuscript

Quantum computing, internet-of-things, solar energy, 3D printing, 5G networks, blockchain, robotics, artificial intelligence… the list goes on, yet we are still using Bronze Age technology – meditation – for consciousness. Genetic cognitive enhancements are the future. The future is now, and the future belongs, as it always has, to visionaries.

This book introduces a practical science for using gene editing to generate sustainable higher states of awareness. Its theory rests on the timeless spiritual principles of the Eastern wisdom traditions. The book concentrates on non-local consciousness and non-local memory, and treats these subjects with a scientific rigor, clarity and simplicity. It considers these subjects from an electrical engineering perspective, which represents a fresh, original approach. The resulting body of work, cognitive neuroengineering, is a transformative technology, and its widespread application can accelerate the awakening of humanity.

J.L. Mee

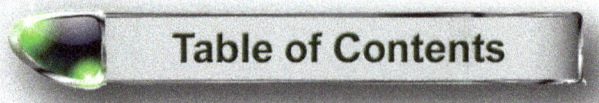
Table of Contents

1. Introduction — 9
Revolutionary breakthroughs in consciousness

2. Cognitive Mechanics — 33
Cardinal laws of awareness

3. Macro memory — 75
Topography and behavior of a conscious being's unconscious mind

4. Cognitive Neurophysics — 93
Non-physical origins of brainwaves

5. Cognitive Neurodynamics — 141
Dynamic interplay between consciousness and neurology

6. Cognitive Neurogenetics — 161
Gene therapies for stable higher states of being

7. Cognitive Ecosystem — 187
Support networks for genetically-enhanced individuals

8. Summary and Conclusions — 199
Global wisdom initiative

9. Epilogue — 213
Extraordinary circumstances leading to discoveries

10. Reference Section — 245
Additional resources

Foreword

The development of genetic cognitive upgrades represents a turning point in the history of human civilization. *Cognitive Engineering* is the first book ever to disclose genetic engineering blueprints for creating a new human genotype with permanently-expanded awareness. The discovery of how to optimize human neurology to enrich cognitive ability has the power to shape the course of history.

Drawing on India's wisdom traditions as well as the work of modern consciousness pioneers such as Dr. Dean Radin, *Cognitive Engineering* unravels the mystery of how non-local consciousness interfaces to physical neurology. The book introduces the world's first mathematics of consciousness – cognitive physics – which expresses the fundamental laws of awareness and its relationship with physical neurology in quantitative terms using scientific equations. Its axioms and formulas reveal cardinal laws governing the dynamic interplay between awareness and the material universe. These laws are conclusively validated by scientific experiments.

Widespread application of this technology can enrich humanity's cognitive potential, opening the way to a golden age of wisdom, and fostering accelerated progress in the arts and sciences. The opportunity to lift the human race into universal self-awareness represents a defining moment in the progress of human civilization.

J.L. Mee

July 13, 2019

Let noble ideas come to us from all directions.
– Rig Veda

Introduction

Introduction

A. Overview

● **Choose Your State**

Startling new information has come to light which fundamentally alters our understanding of consciousness. Permanent higher states of being can be achieved through the application of dramatic new advances in awareness science and genetic engineering. These breakthroughs have created an historic new methodology for achieving stable states of expanded awareness.

In this century alone, millions of people have spent billions of hours trying to free themselves from compulsive thinking in meditation. In the near future, you may be able to permanently master compulsive thinking just by taking a pill. In fact, in the not-too-distant future, you may even be able to choose the flavor of meditative state you want – Zen, Vipassana, TM, Mindfulness, Concentration, Loving Kindness – and take a gene therapy pill to optimize your neurology for the state you selected.

Normal Human Awareness

How would it feel to be super-aware? What if you had twice your current awareness? What if you could experience stable, permanent higher states of being, expanded present time awareness, elevated intelligence, and greater overall life force? Would you enjoy life more?

These dramatic advances in consciousness engineering offer exciting possibilities for today's awakened young adults. People in their twenties and thirties today could conceivably spend most of their lives in permanent higher states of being. This new generation of superconscious individuals can form the vanguard of enlightened leadership in the arts and sciences.

Genetically-Enhanced Awareness

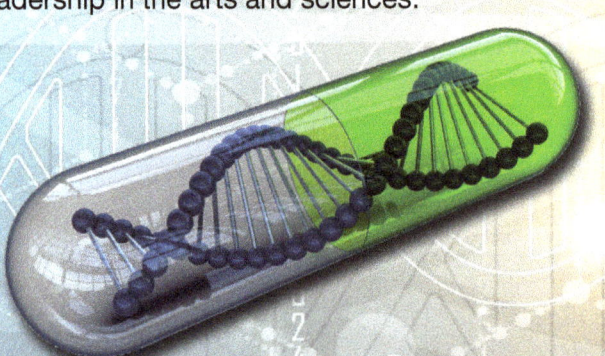

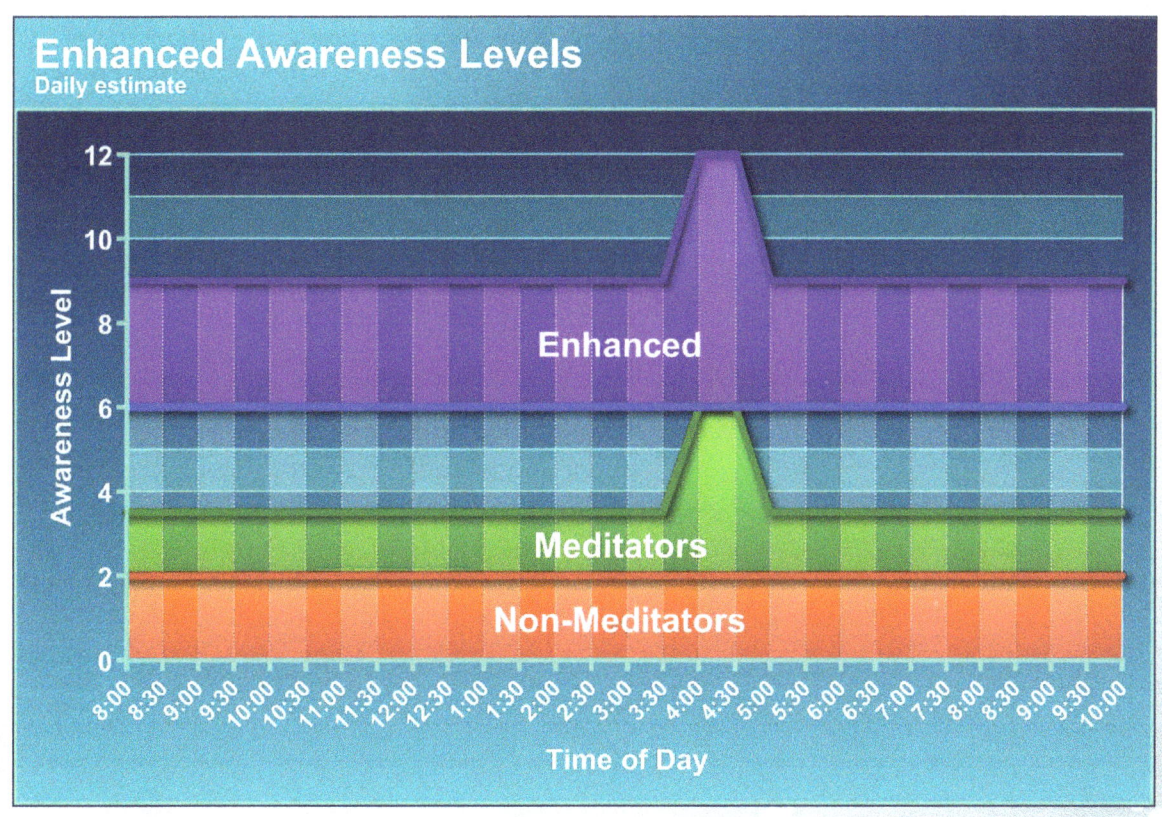

Benefits

Meditation has been scientifically proven to deliver a cornucopia of mental and physical benefits.

The consciousness engineering technology this book introduces has the power to raise a person's baseline level of awareness to a meditative state…so instead of spending half an hour a day in an expanded state, they spend all day.

What happens when a person spends all day in a meditative state instead of 30 minutes? Ten hours in meditation generates 20 times the benefits of half an hour.

That adds up to 20 times more benefits in self-awareness, mindfulness, optimism, happiness, well-being, physical health, learning, mental acuity, creativity, compassion, trust, empathy, mental health, and emotional intelligence.

Consciousness engineering also has the flexibility to provide customized levels of gain based on individual needs. For instance, living in exalted states of being will not be practical or even desirable for millions of people who are busily involved with their families and careers. The technology can be dialed down to generate milder gains in awareness and cognitive performance for this audience. Entry level cognitive upgrades can deliver significant benefits across a wide cross section of our population, and they will represent a key channel for distributing the technology.

The Arc of History

Humans have sought the fountainhead of higher awareness for thousands of years. However, due to the inherent challenges in attaining advanced states of being, the fruits of higher knowledge have been limited to a relatively small group of people with years of dedicated practice.

The opportunity to open up stable higher states of being to the populace at large represents a defining moment in the progress of human civilization. Such a monumental advance would empower a new generation to reach up and bend the arc of history in the direction of more freedom, greater wisdom, and more compassion.

Science and neuroscience are making landmark discoveries at a rate unknown in our lifetime or across history. A relentless parade of new scientific developments is unfolding on many fronts. The pace of advance is accelerating, and there are many rapidly evolving, potentially transformative technologies on the horizon. This book identifies revolutionary new developments which could have a massive impact on humanity's collective wisdom between now and 2050. It also discusses how these technical breakthroughs could change our world and recommends ways to capitalize on them.

Cognitive Engineering presents hitherto unknown laws of nature which enable scientists to treat states of awareness quantitatively with mathematics. This advance represents a cardinal paradigm shift in our understanding of consciousness. Widespread application of these discoveries around the globe can change the course of human civilization.

The world today is shifting into a knowledge economy where cognitive capital is the prime currency.

In the 21st Century, the ability to marshal this precious resource for creative problem solving will determine the economic mobility of individuals and the wealth of nations.

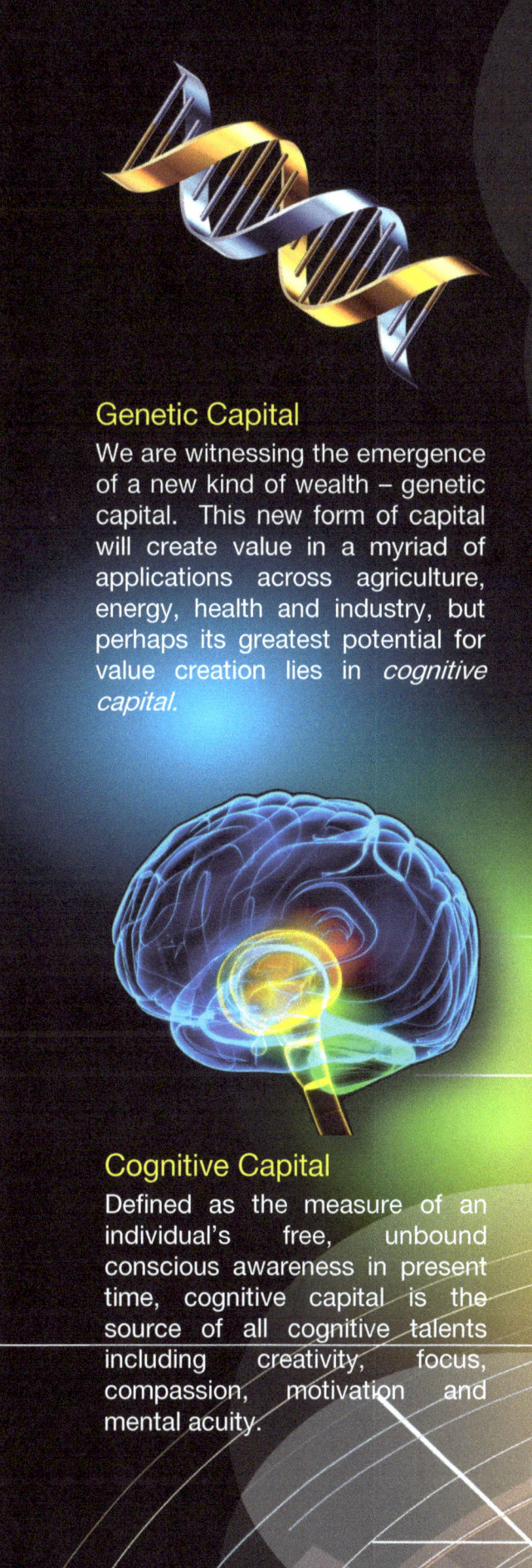

Genetic Capital

We are witnessing the emergence of a new kind of wealth – genetic capital. This new form of capital will create value in a myriad of applications across agriculture, energy, health and industry, but perhaps its greatest potential for value creation lies in *cognitive capital*.

Cognitive Capital

Defined as the measure of an individual's free, unbound conscious awareness in present time, cognitive capital is the source of all cognitive talents including creativity, focus, compassion, motivation and mental acuity.

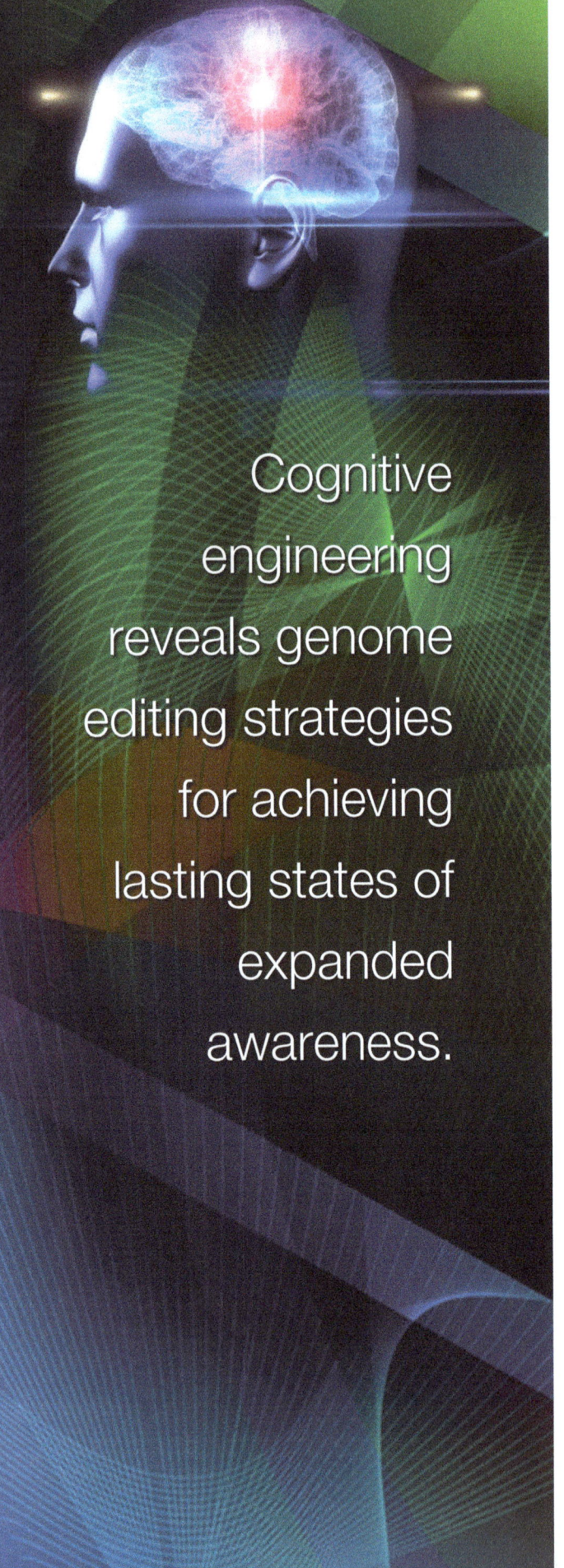

Cognitive engineering reveals genome editing strategies for achieving lasting states of expanded awareness.

● Unlocking Potential

New research has revealed vast reserves of human cognitive potential buried beneath layers of unconscious memories, thoughts and behavior programs.

A revolutionary technology for unlocking this potential to generate stable expanded states of awareness, called *cognitive engineering*, has arisen at the confluence of two new branches of science, *cognitive physics* and *genetic engineering*.

The principles of cognitive physics are anchored in the timeless truths of India's great wisdom traditions, which form the bedrock for a new 21st Century mathematics of consciousness. These simple and elegant formulas yield a new awareness science for the Third Millennium.

Cognitive physics reveals seminal new discoveries about the nature of awareness and its interactions with human neurology. These insights yield genetic engineering strategies for optimizing the brain's neurological substrates for cognition.

Biotechnology enterprises can leverage this research to develop genetic engineering solutions for elevating human consciousness to sustainable higher states of being.

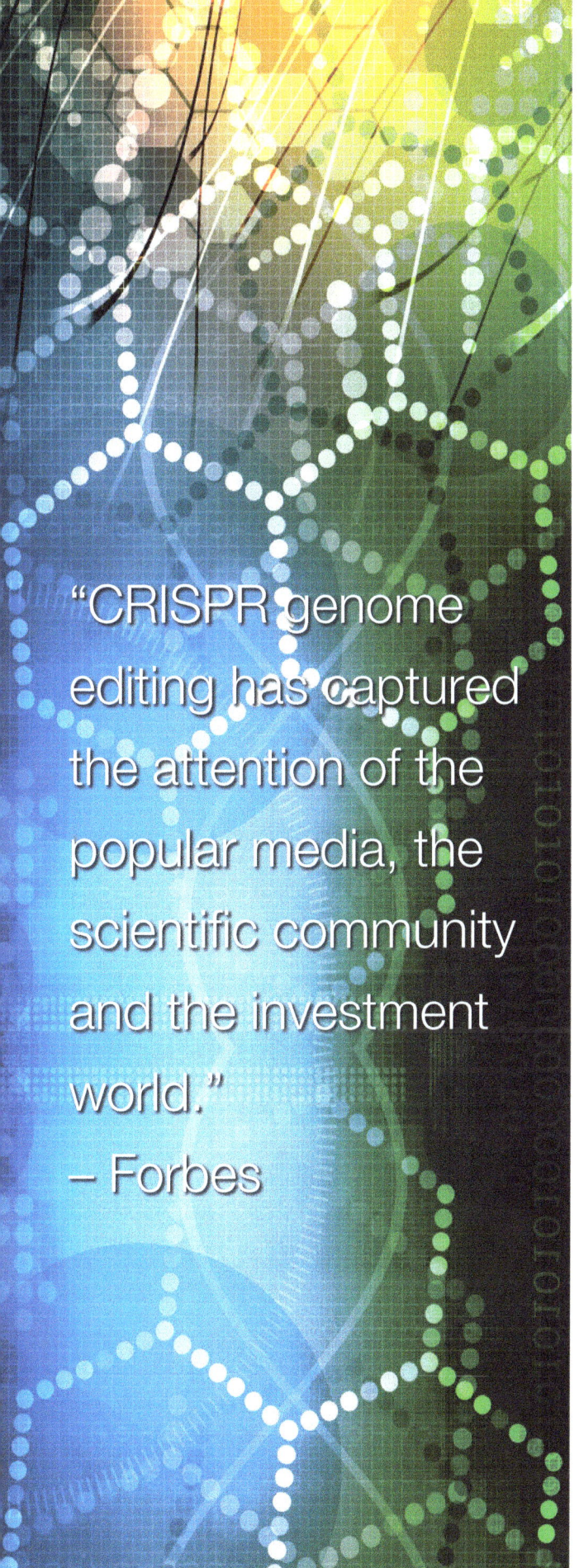

"CRISPR genome editing has captured the attention of the popular media, the scientific community and the investment world."
– Forbes

Genetic Technology Trends

The human race is standing at the threshold of a Genetic Age. Genetic engineering can enhance cognitive faculties such as mental acuity, awareness and intelligence, and physical attributes such as appearance, strength, and agility. Investment in gene therapy is booming.

According to The Alliance for Regenerative Medicine, public and private genetic engineering companies raised $22 billion globally in the last three years.

The previously-unfathomable mechanisms of gene interaction have been penetrated by a revolutionary laboratory technique called CRISPR. This development is a seminal achievement in the history of biological research. Its impact on human civilization is being compared to the steam engine, which gave rise to the Industrial Revolution, and the transistor, which spawned the Information Age.

The Genetic Age dawning today will experience a power growth curve similar to the one the Information Age has enjoyed over the last six decades. The computer industry was in a very primitive stage when it began sixty years ago, but investors who were far-sighted enough to grasp its enormous potential built great fortunes.

CRISPR Revolution

"The discovery of the century."
— Bloomberg

"DNA editing will remake the world."
— Wired

"The potential is enormous."
— TIME

"This revolutionary gene editing
Tool could change the world."
— NBC News

"The gene-editing tool's potential to
upend science is dizzying."
— Vox

"CRISPR really will change the
world forever."
— ScienceAlert

Extraordinary Opportunity

"Talk to any biologist right now and you will hear a level of excitement that comes only from the emergence of something truly groundbreaking…
If the evolution from giant mainframes to personal computers forever changed technology, CRISPR promises to do something similar for genetics…
The potential is enormous…"
– TIME Magazine

CRISPR

- Allows scientists to edit DNA like programming a computer
- Tremendous improvement in DNA editing speed, ease, cost and accuracy
- Monumental landmark in the history of biological research
- Impact on human civilization being compared to the steam engine and the transistor

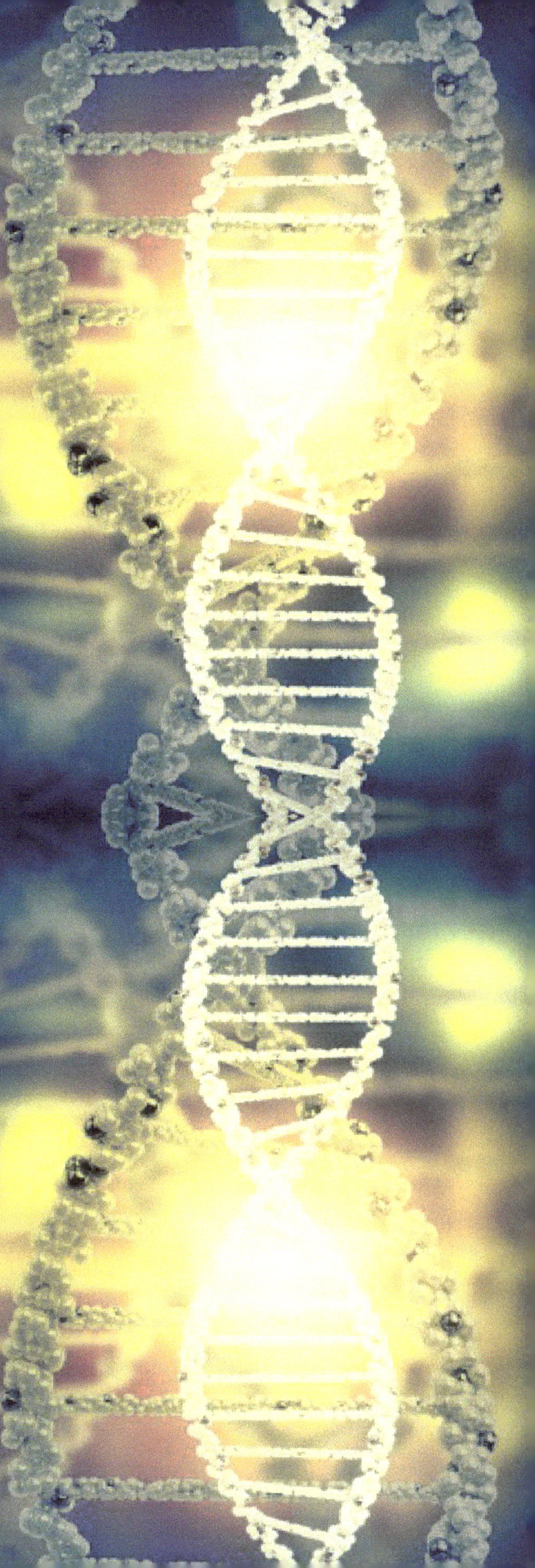

● Opportunity

Today, civilization is perched on the cusp of a new era...the dawn of The Genetic Age. To appreciate the economic potential of this new age, we have only to look back at the last era for evidence. How much wealth was created over the last fifty years in the Information Age?

Never are opportunities greater than at times like this of great transition. The invention of the transistor revolutionized many different industries. Computing, television, avionics, process control, appliances, automobiles and many other fields were changed.

An entirely new industry is emerging to leverage the power of CRISPR to transform human cognitive ability. This industry will contain many different companies, products and services. If history repeats itself, the early players in this new field will become the market leaders.

Human neurology can be optimized for consciousness.

● Scope

In the same way transistors were applied across many different fields, genome editing will be used in many diverse human applications. Genetic engineering can be applied to improve human abilities in four ways:

Cognitive: Cognitive faculties such as self-awareness, mental acuity, and intelligence.

Anatomy: Physical attributes such as strength, agility, beauty, grace and stamina.

Longevity: Elements of long life including wellness, immunity and metabolism.

Talent: Mental, physical and emotional talents.

The scope of this book is limited to the first category, *cognitive enhancement*.

The Computer Age

In December, 1947, three young engineers at Bell Labs in Murray Hill, New Jersey – John Bardeen, Walter Brattain and William Shockley – demonstrated a history-changing invention to the company's management. They called their new device a transistor, and it ushered in the Information Age. Today, humanity has reached a new turning point in history which is just as monumental as this pivotal discovery.

The human race is standing at the threshold of a Genetic Age. New genetic engineering tools are emerging which represent breakthroughs so fundamental they are being compared to the invention of the transistor.

The computer industry was in a very primitive stage when it began, but those who were far-sighted enough to grasp its enormous potential profited handsomely from their investments and built great fortunes. Who could have guessed the humble beginnings of the Information Age would eventually culminate in supercomputers, smart phones and the global internet? Moreover, who could have imagined a democratization of information that would empower people the world over with instant access to virtually all human knowledge?

Progress of Human Civilization

Table 1 shows the master organizing principles underlying three stages of human civilization: the Industrial Age, the Computer Age, and the Genetic Age.

The emergence of current genome editing technologies has been widely compared to the advent of the steam engine and the transistor.

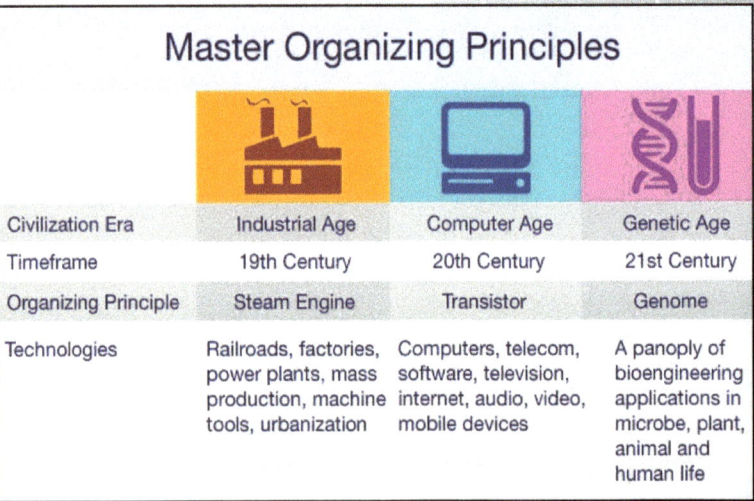

Civilization Era	Industrial Age	Computer Age	Genetic Age
Timeframe	19th Century	20th Century	21st Century
Organizing Principle	Steam Engine	Transistor	Genome
Technologies	Railroads, factories, power plants, mass production, machine tools, urbanization	Computers, telecom, software, television, internet, audio, video, mobile devices	A panoply of bioengineering applications in microbe, plant, animal and human life

– Table 1 –

The Genetic Age

The Genetic Age which is dawning today will experience a growth power curve similar to the one the Information Age has enjoyed over the last six decades. To peer into that future for a moment, imagine an engineering organization ten times the size of Intel (at its peak, 80,000 employees). Instead of implementing computer designs in silicon, these engineers implement body designs in DNA.

Assume these 800,000 genetic engineers have a decade or two to make progress. What kind of enhancements do you think they could create? A scientific powerhouse like this could eventually clear away pain, suffering, disease, languor – every human imperfection – and replace them with strength, agility, health, grace and super intelligence. The result will be top-of-the-line equipment. Living in one of these bodies will be like driving a Bentley. How would you like to have one?

Imagine there is a comparable-sized organization that develops consciousness science in parallel with the advances in genetic engineering. With ten decades of progress on the consciousness front, what could happen? Could these engineers come up with brain designs that offered greater intelligence and awareness? How about five or ten times the present level of human consciousness and intelligence?

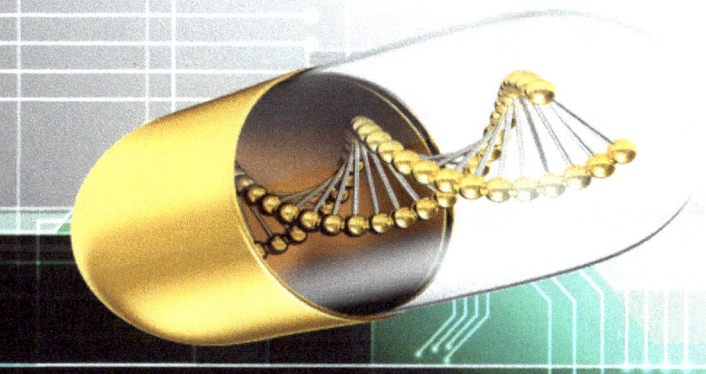

Gene therapy can be used in hundreds of ways to improve the human body, including longevity, health, wellness, mental acuity, intelligence, beauty, stamina, strength, immunity and resiliency, just to name a few.

"The technology's possibilities are staggering."
– Fortune

Cognitive Engineering

● The Riddle

Have you ever experienced a heightened state of awareness, only to have it fade? Did you ever wonder where the extra awareness came from? Did you ever wonder where it went?

When most people experience higher states, the states are transitory; they come and go. This book solves an age old riddle: "Where does expanded awareness come from, and where does it go?" Solving the riddle opens the door to a new science of consciousness which provides a roadmap that can lead humanity to permanent higher states of being.

● Sustainable Awareness

We know awareness is scalable; the age-old question is how to make it sustainable. This book presents a radical new approach to achieving permanent gains.

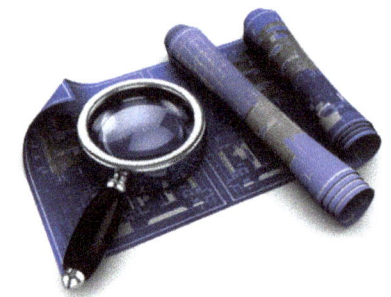

Cognitive Engineering discloses the blueprints for genetically engineering human neurology to support persistent expanded states of consciousness. These blueprints are based on revolutionary, laboratory-proven discoveries which reveal – for the first time in history – how embodied conscious beings communicate with their physical neurology. This new branch of human knowledge, called cognitive physics, provides the scientific know-how for genetically optimizing neurology to support stable higher levels of awareness.

Societies, groups and nations which embrace this disruptive new technology can harness its vast potential to drive exponential growth in cognitive capital. Its widespread application can enrich humankind's cognitive resources around the world and foster accelerated progress in the arts and sciences. When realized, this achievement will arguably constitute one of the greatest advances in human history, leading to a mass uplifting of consciousness on the planet, and opening a golden age of wisdom for humanity.

Cognitive Engineering

B. Benefits

The influence of cognitively-enhanced individuals will spread through their families, communities and workplaces, ultimately benefitting humanity as a whole.

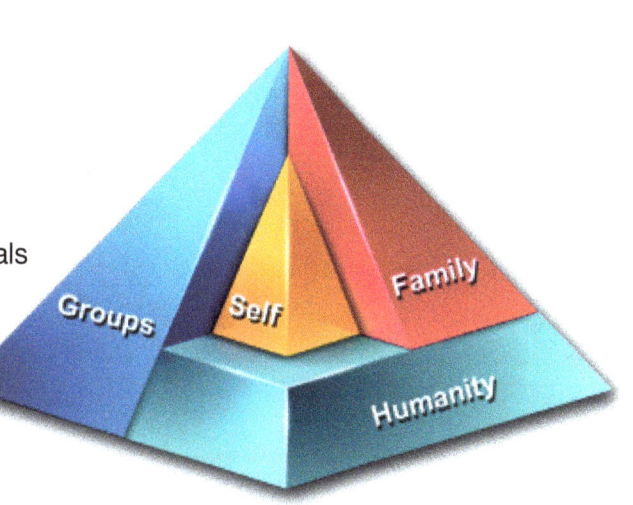

1. Benefits to Individuals

Living in continuous awareness-of-awareness will enable people to effortlessly remain centered in higher consciousness as they go through life. This enriched cognitive capital will generate dividends in inner peace and tranquility, greater mental clarity and acuity, and enhanced mindfulness in the present moment.

Cognitive engineering can produce wise, centered and resourceful individuals who are free from compulsive thinking. These superaware people will enjoy greater personal sovereignty and freedom, along with an expanded perspective on life heretofore unmatched in the mainstream of human history.

• *Personal Benefits:* Cognitive engineering releases a person's awareness from the past, restoring greater conscious awareness into the present. This delivers cognitive advantages similar to meditation. Meditation has been widely tested and found to produce improvements in creativity, emotional balance, mental acuity, concentration, motivation, learning, mindfulness, optimism, happiness, well-being, physical health, compassion, trust, empathy, and emotional intelligence.

• *Economic Value:* Cognitively-enhanced men and women make ideal employees for firms of all sizes competing in the knowledge economy. They have greater cognitive capital to apply to developing creative solutions to business problems, as well as higher morale and superior emotional balance. Hence, genetic upgrades have an economic value, representing an investment similar to higher education which can pay dividends throughout a person's professional life in the form of career advancement.

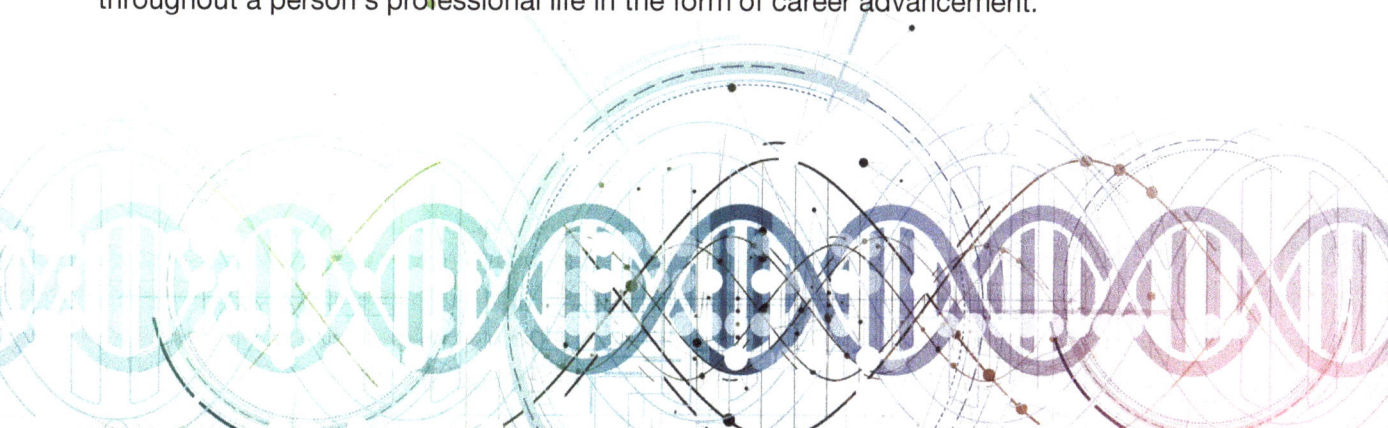

Transformational Results

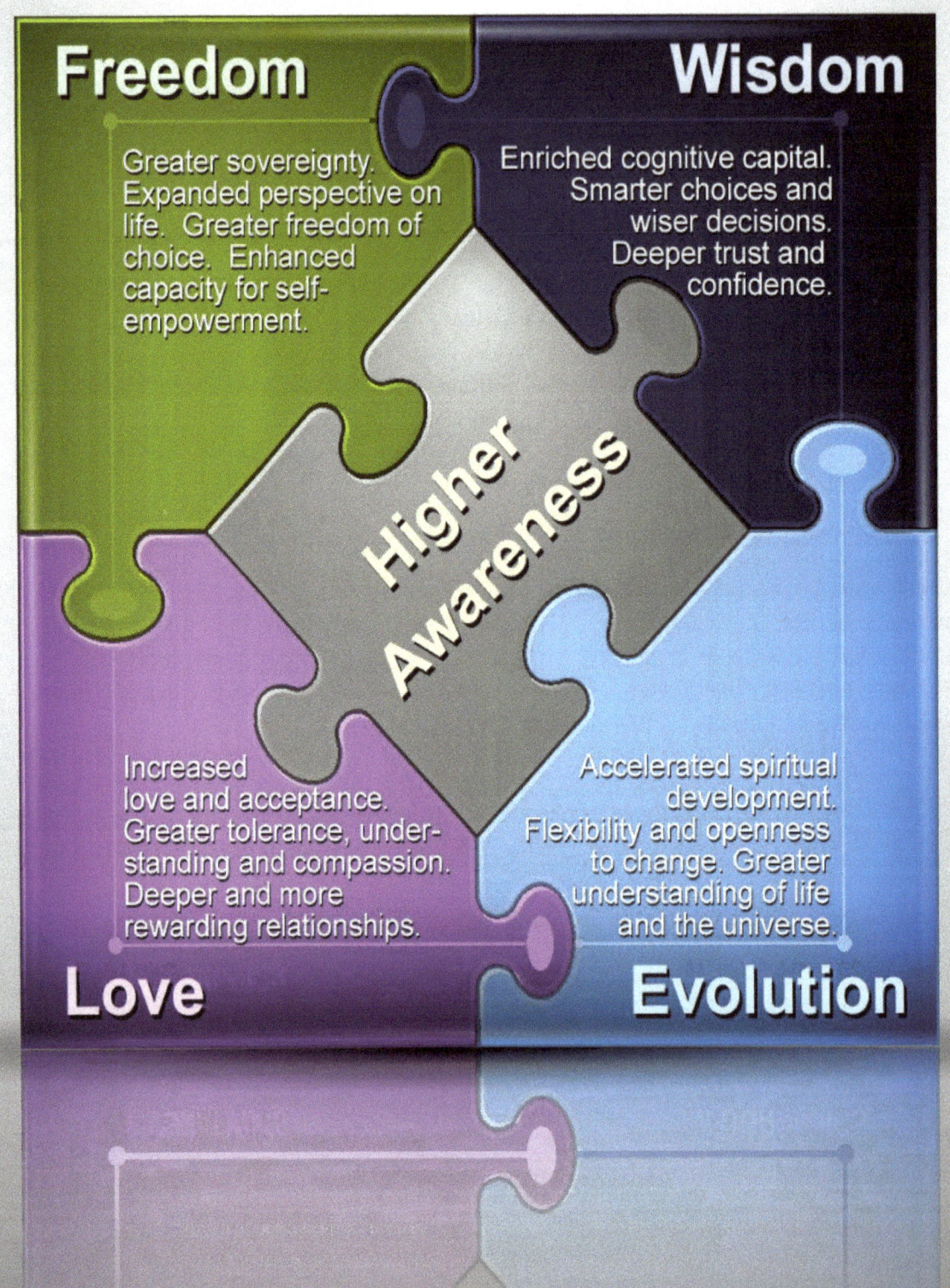

The advantages of elevating a person's level of conscious awareness for an entire forthcoming lifetime are gargantuan. This upgrade energizes a person's limitless higher self-awareness, which can deliver a wealth of immediate and long-term dividends, such as:

Expanded sense of self

- Trust and confidence in the future
- Open to exciting new possibilities

Greater sovereignty

- Expanded perspective on life and greater freedom of choice
- Enhanced capacity for self-empowerment and understanding

Inner peace and tranquility

- Mindful living in the present moment
- Initiate and engage in positive thoughts and behaviors

Accelerated consciousness development

- Heightened awareness and intuition
- Greater understanding of the mysteries of life and the universe

Flexibility and openness to change

- Facilitate lasting paradigm shifts
- Release limitations to a more authentic self
- Make smarter choices and wiser decisions

Increased love and acceptance

- Create deeper and more rewarding relationships
- Greater tolerance, understanding and compassion

2. Benefits to Families

The influence of cognitively-enhanced individuals will spread through their families, communities and workplaces, ultimately benefitting humanity as a whole.

Expanded consciousness fosters harmony in human relationships. It enriches the spiritual dimension of families, bringing more awareness to family dynamics, and creating deeper and more meaningful relationships between spouses.

Gene therapies for higher consciousness will only be made available to adults over 21 years of age. However, these therapies for greater awareness can benefit children indirectly in the form of more conscious parenting by adults.

Children can also look forward to lives of enriched awareness when they become adults. Living superaware lives can fast track their personal evolution towards greater wisdom.

Today's younger generations respond to vigorous new concepts that have a dynamic vision and a scientific luster. The exciting results of cognitive upgrades can counterbalance the mesmerizing effect of affluence and luxury on young adults, yielding a more holistic approach to life.

3. Benefits to Groups

Widespread application of cognitive engineering will tend to intensify public awareness of conscious organizations and groups, raising engagement and participation in their services and activities.

Charities: Socially-conscious philanthropic causes and charities may experience tailwinds from publicity generated by historic advances in consciousness engineering, helping them to attract more donors, volunteers and supporters.

Communities: Greater peace of mind and emotional balance will reduce conflicts individually and societally and improve community coherence. Higher awareness of human interdependency may energize participation at local community and cultural centers.

Workplaces: Enriched cognitive capital in their workforces can help companies achieve greater success in the new knowledge economy. Increased happiness and positive emotional states will also engender higher employee wellness.

4. Benefits to Humanity

A global shift towards higher consciousness will help people to realize their interconnectedness with the human community and nature. Heightened cognitive abilities can be applied to solving some of the planet's biggest challenges and problems.

Elevating world awareness of higher states of being will cause policy makers at every level to give greater consideration to the impact of their decisions on future generations.

Humanity's growth into higher awareness will tend to defuse irresponsible short-range public and corporate policies which emphasize near-term results at the expense of long-range sustainability. As thousands of key decision makers start to see themselves as inheritors of the future, they will forge more enlightened policies which reflect a more responsible stewardship of the planet.

C. New Science

Before we can genetically optimize the brain to support stable states of higher consciousness, we must understand the true nature of awareness and the mechanics of how it interacts with neurology. Accordingly, most of the book is devoted to these two essential subjects. Once they are well understood, they illuminate genetic engineering pathways for achieving the goal.

New Branch of Science

The study of wisdom in India is over 3,000 years old, but electrical engineering is a very recent development in human history. It has been widely used only in the last 100 years, which is just 3% of the last 3,000 years. This book considers the nature of awareness from an electrical engineering perspective. Since electrical engineering is new, this line of thinking has not been pursued before. Hence, this book contains a body of original work.

Since the end product of a conscious being's mental activity is brainwaves, examining conscious beings from the viewpoint of electrical engineering proves to be auspicious. If conscious beings produce electromagnetic brainwaves, then electromagnetic waves are a good entry point into understanding conscious beings. Nothing explains electromagnetism better than electrical engineering.

Once we understand brainwaves, we can work backwards towards their source to uncover the mechanics of how conscious beings produce these waves. We discover that electromagnetic waves are a key element in the interface zone between conscious beings and neural networks in the brain. This insight lays the foundation for a new branch of human knowledge.

Science can be pictured as a tree with many branches, such as astronomy, botany and chemistry. This book adds a new branch to the tree of science, cognitive physics, which covers the study of awareness and its interactions with neurology.

Cognitive Physics

Cognitive physics expresses the fundamental laws of awareness and its relationship with physical neurology in quantitative terms using scientific equations. This breakthrough represents a historic milestone in understanding how mind and matter interact with one another. Cognitive physics comprises five new interconnected sciences which work together to create a revolutionary strategy for raising human awareness through genetic engineering.

Cognitive Mechanics

Cognitive mechanics applies scientific methods to the study of awareness and the way it interacts with the physical universe. Its axioms and formulas define the primary states of awareness and its behavior, locality, mobility, density, and periodicity. They reveal cardinal laws governing the relationship between consciousness and the material universe.

The resulting body of knowledge enables scientific measurement and classification of awareness states.

Cognitive Neurophysics

Cognitive neurophysics explains how information in a conscious being's non-physical memories is propagated into its physical neurology. It reveals how the brain's cellular-level electromagnetic fields serve as the interface zone between consciousness and physiology, and explains the mechanics of how awareness and matter interact in this zone.

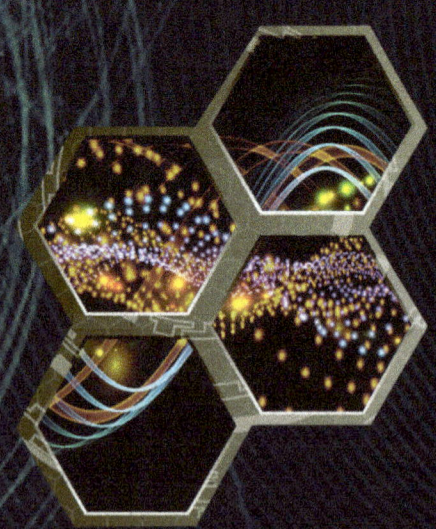

Macro Memory Science

This discipline provides a scientific framework which explains the working mechanics of a conscious being's non-physical memories. It clarifies exactly how non-physical memories travel with the conscious being over multiple lifetimes, and demonstrates how these memories can generate brainwaves and behavior. This breakthrough represents a fundamental paradigm shift and a turning point in the history of neuroscience.

Macro Memory

Cognitive Neurodynamics

Cognitive neurodynamics establishes mathematical equations which explain the relationships between conscious awareness levels and brainwave electrodynamics.

Cognitive Neuro-Dynamics

Cognitive Neurogenetics

Cognitive neurogenetics defines a strategy for developing gene therapies for optimizing human neurology to support permanent higher states of being. It uses insights from cognitive neurodynamics to modify certain neural pathways to restore higher levels of awareness to the conscious being.

Cognitive Neuro-Genetics

Cognitive Mechanics

Awareness Potential

Each of us is inherently capable of higher states of being. A conscious being's awareness potential is considerably greater than most people imagine. A peak experience in someone's current life might bring them to an awareness level 5 times greater than their normal consciousness. From this perspective, they might estimate their full potential to be 15 to 20 times greater than their average level of awareness.

In reality, our actual capacity for beingness is over 100 times greater than our usual level of consciousness.

Original State

There is a growing awareness in the West of the non-local dimension of consciousness which exists outside of space and time. Eastern traditions teach that we have been alive for many eons, and will be alive for many more.

The Vedic texts posit that in our original state, we were created out of Godhead long ago as beings of pure awareness, and over the course of eons, we descended from this angelic state into our present human form. The sweeping dimensions of this philosophy may be open to debate, but the point is that we must not confuse our present limitations with our actual potential.

Awareness Potential

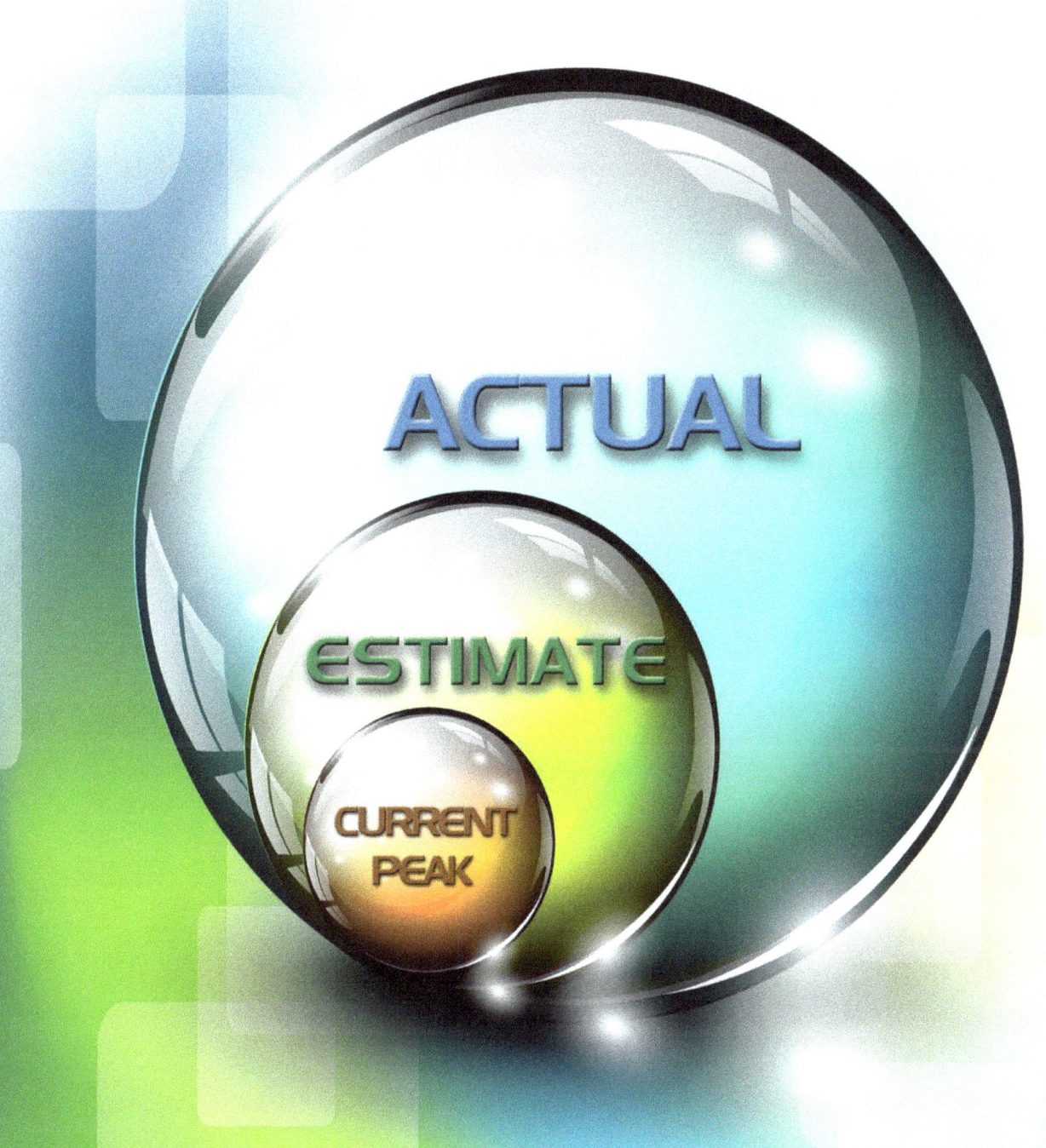

● **Awareness Levels**

What would it look like to be fully conscious? To answer this, the diagram on the facing page illustrates an awareness scale with 100 levels. At the top, the person is fully conscious, and at the bottom, completely unconscious.

The lower states reflect denser forms of awareness, such as sleep, while the higher levels contain lighter, super-conscious states. The top tiers represent angelic states that are rarely found on Earth.

The numbering of the levels on the vertical axis reflects the density of awareness, so the lightest states are at levels 1-20, and the densest states are at levels 80-99.

The green bar identifies a human being's normal waking state, and the blue bar marks the state of sleep. As illustrated, the difference between a human being's normal state (the green bar) and their sleeping state (the blue bar) is almost trivial compared to their full potential for awareness.

Most human beings are only shadows of their true selves. Moving into our larger identity enables us to make wiser choices for our future.

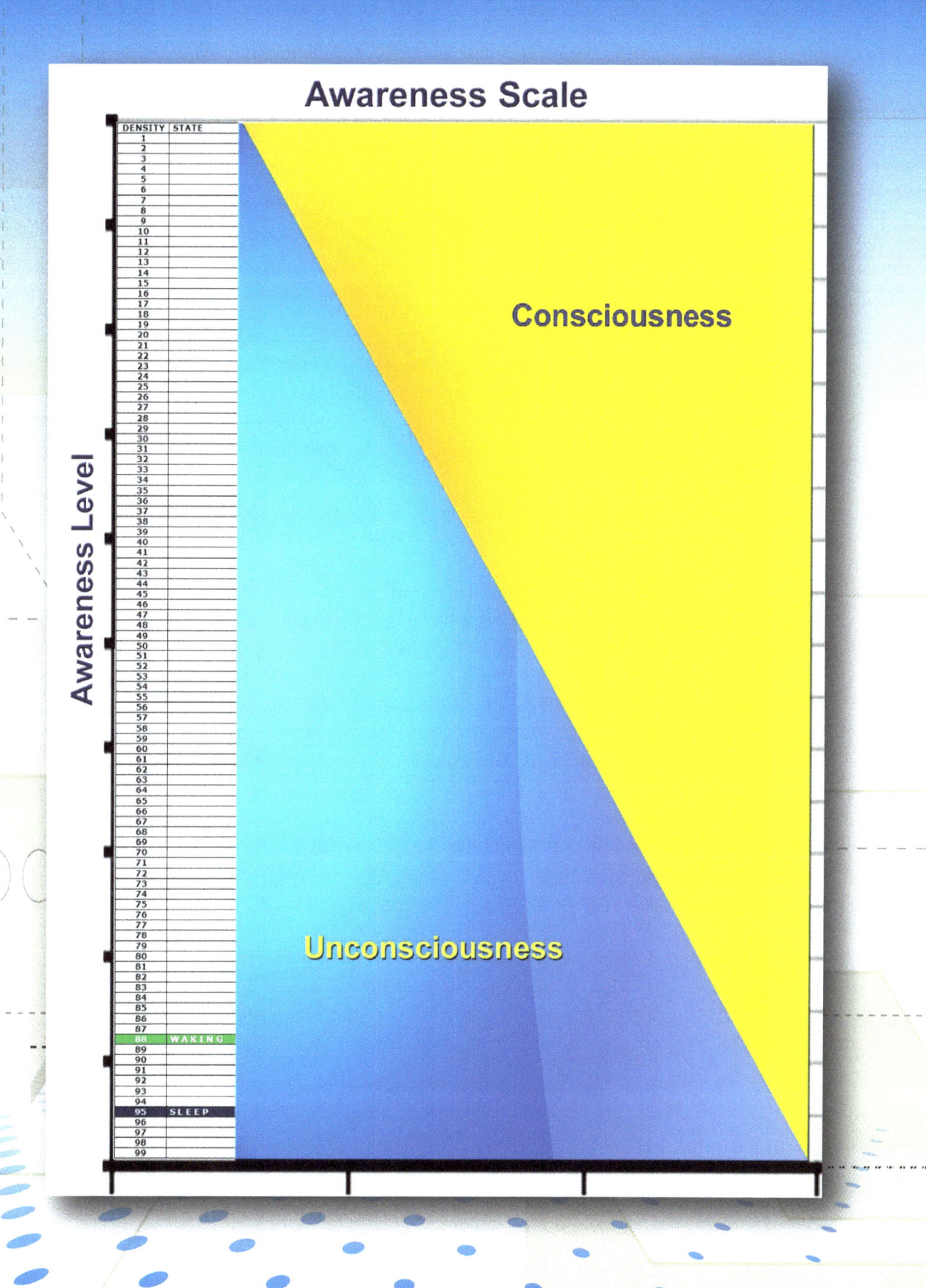

⬤ Awareness Metrics

If we have such enormous potential for awareness, where is it? Where is all this extra awareness hiding? Why isn't everyone superaware?

The answer is the extra awareness is unconscious, but the situation is worse than you think. A common misconception is that although human beings are mostly unconscious, if we were somehow catapulted into our full potential, we would be supercharged versions of who we think we are now. This is untrue.

Actually, at our full potential, we are beings of such magnificent radiance that the human imagination cannot even begin to comprehend it. It is like going from dim bulb to miniature sun.

That's the good news. The flip side of the story is the pile of unconsciousness is much higher than you think. The extra awareness we don't know we have is being soaked up by extra unconsciousness we also don't realize we have.

This see-saw-like interplay between consciousness and unconsciousness is a critical relationship, and understanding it is essential for developing sustainable higher states of being.

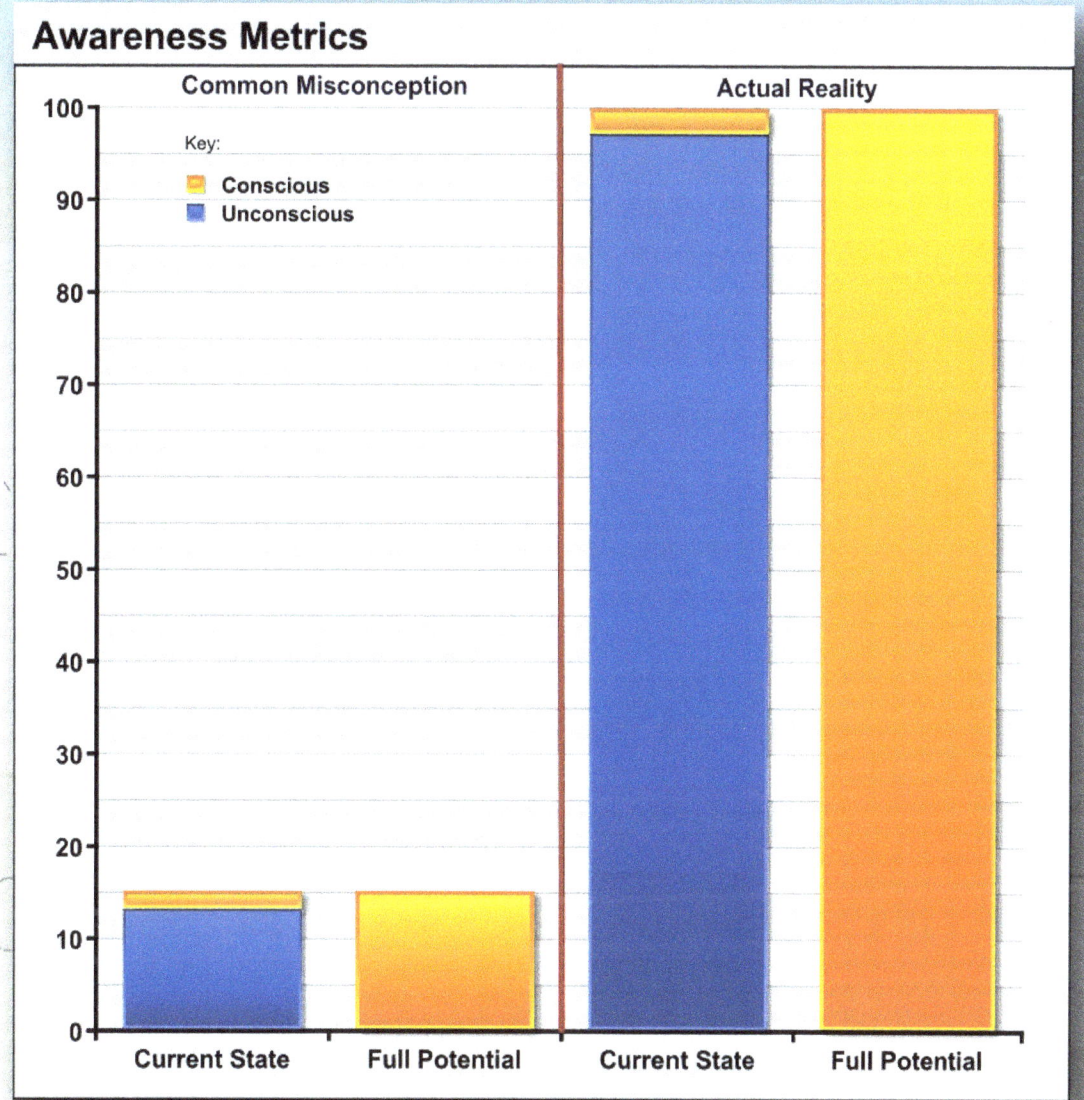

Cognitive Engineering

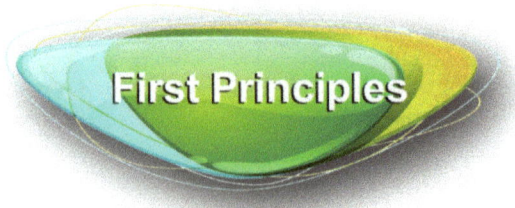

● Introduction

Cognitive mechanics applies scientific methods to the study of consciousness and its relationship to the physical universe. Its axioms and formulas define the primary states of awareness and its behavior, locality, mobility, density, and periodicity. They reveal fundamental laws governing the relationship between consciousness and the material universe. The resulting body of knowledge:

- Enables scientific measurement and classification of awareness states;
- Enumerates the characteristics of higher states of being, contributing to our understanding of life in the universe;
- Provides the foundation for macro memory science; the essential element in establishing the relationship between unconscious memories and brainwaves.

Axiom 0 – Identity
A conscious being is a unit of identity.

A conscious being is a discrete unit of identity comprised of awareness. Awareness itself is composed of fundamental units called *awareness points*. Conscious beings are comprised of awareness points in much the same way as light is made of photons, or matter is formed out of atoms.

Axiom 1 – Awareness Dual State
Awareness has two states: free and applied.

Photons can change their states to exhibit the behavior of a particle or a wave. Similarly, awareness points can behave like non-physical particles or waves. This behavior results in two kinds of awareness; *free awareness*, which is conscious, and arises from particle states, and *applied awareness*, which is unconscious, and arises from wave states.

Free awareness: *Free awareness is conscious and formless. It is stationary, changeless, and exists outside of time and space.*

In *Conscious Universe* (Harper Collins, 1997), Dr. Dean Radin uses the term "non-local consciousness" to describe this type of free awareness, and explains that it has no location in space or time.

Free awareness contains no mass, energy, space, amplitude or wavelength.

The idea of free awareness is not new. The Vedas observed in antiquity that pure conscious awareness has no form. Had they known about electromagnetic waves, the Vedas might have concluded that pure consciousness also has no wavelength (since wavelength is a form). However, the electromagnetic spectrum was not known until European scientists Michael Faraday and James Clerk Maxwell discovered it in the mid-19th Century. With this knowledge, we can deduce that pure consciousness has no wavelength, and this idea is one of the pillars of cognitive engineering.

Applied awareness: *Applied awareness is unconscious and has a form. It is located in time and changes its characteristics in time.* Its fluctuations over time are called oscillations, and they form non-physical waves called *memory waves*.

Awareness changes state from *free* to *applied* when free awareness is applied to a form, such as a memory. The applied awareness permeates and animates the form. In this process, the free awareness becomes unconscious.

When applied awareness is removed from a form, such as a memory, the awareness reverts back to its original free and conscious state.

Applied awareness powers the functions of memory and physical life. It could also be called "mind" or "mental energy."

Note: For nomenclature definitions of "conscious" and "unconscious," refer to the Semantics chapter in the Reference section at the back of the book.

Axiom 2 – Condensation
The living universe is a condensed form of consciousness.

When consciousness first condenses, it forms memory. Memory is the dimensional interface between awareness and the physical universe; the essential foundation for embodied life.

Without memory, we could not learn language, we could not think or form an identity, and we could not recognize people, places or things. Every time we learned something, like how to walk or open a door, we would forget it.

Memory is the gateway for conscious beings into life in the physical universe of space, energy and matter.

Memory is applied awareness, and represents awareness in an unconscious form. As it condenses, it first occupies *space,* and causes fluctuations in the quantum fields of space. As it condenses further, it occupies *energy*, and generates electromagnetic waves. These waves affect *matter*, specifically flowing electrons in the brain, which drive the neural network activity that powers the functions of physical life.

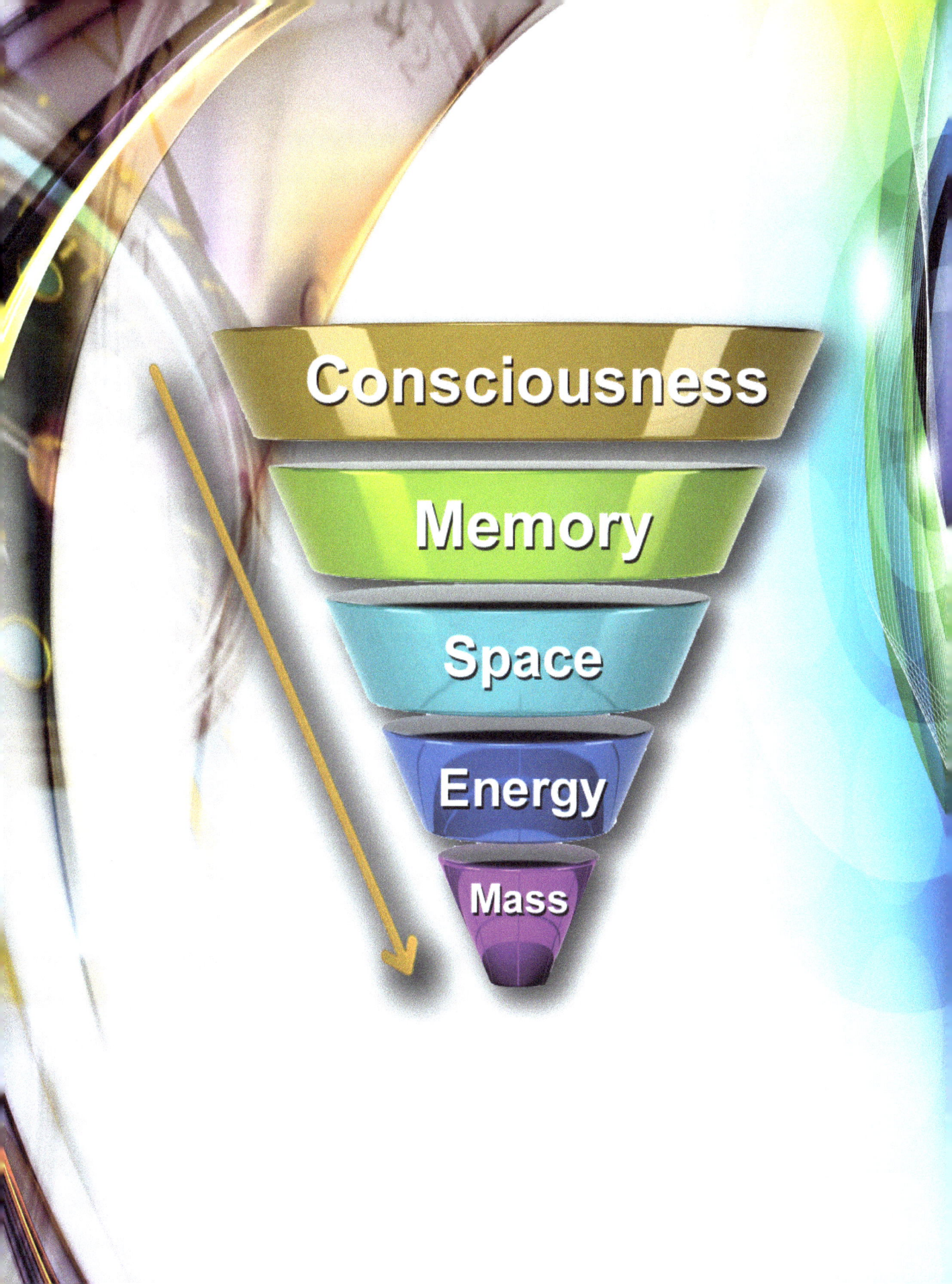

Vedic explanations of a conscious being's relationship to the physical universe are summarized in the following two axioms and corollaries.

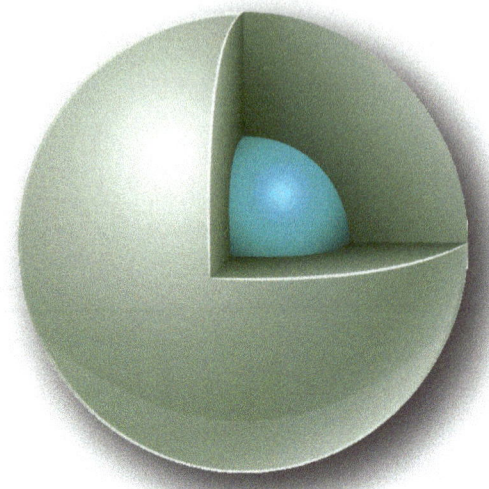

 ### Axiom 3 – Original State
In its original state, a conscious being has no location in space.

The physical universe of matter, energy and space, shown in blue, is surrounded by the awareness universe of consciousness, shown in silver. In our original form, conscious beings exist in the silver sphere outside of the physical universe.

 ### Axiom 4 – Locality
A conscious being occupies a location in space through incarnation.

Conscious beings enter the physical plane through incarnation. Awareness assumes a location in space to animate neurology and create life.

 ### Corollary 1 – Non-Locality
Between embodiments, conscious beings have no location in space.

Bodyless conscious beings reside in the awareness universe, and are not located in the physical world.

Corollary 2 – Mobility
Physical distance is not a factor in the interlife.

Conscious beings who are in between lives reside in the awareness universe surrounding the physical world.

One of the cornerstones of physics is the Law of Conservation of Mass, developed by Antoine Lavoisier in 1748. This law states that matter cannot be created or destroyed; only changed from one state to another. A similar law operates in the consciousness universe.

 ### Axiom 3 – Law of Conservation of Awareness
Awareness cannot be created or destroyed, only changed from one form to another.

The three states of matter are solid, liquid and gas. The two states of awareness are conscious and unconscious.

As shown in the graph, the measurements for consciousness and unconsciousness are reciprocals of each other; as one increases, the other decreases. However, an individual's total amount of awareness remains constant, reflecting a law of conservation.

Law of Conservation

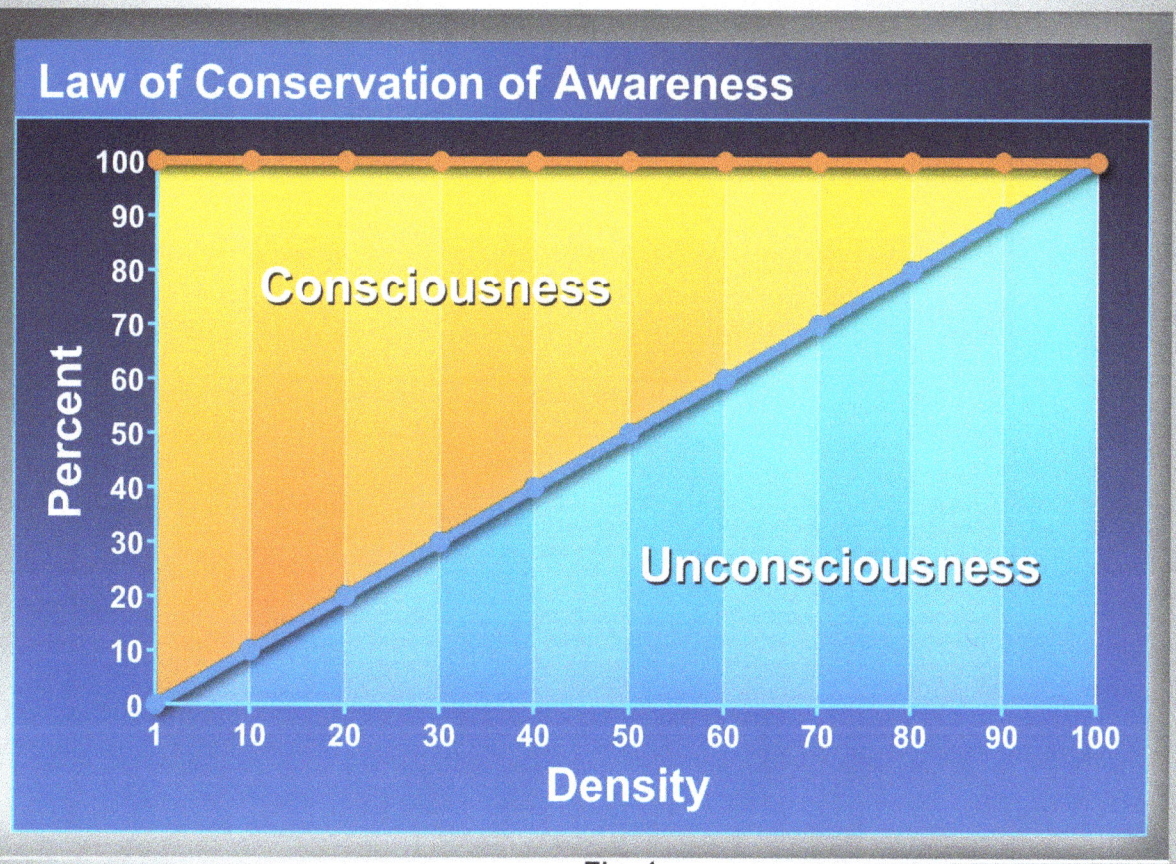

- Fig. 1 -

Cognitive Engineering

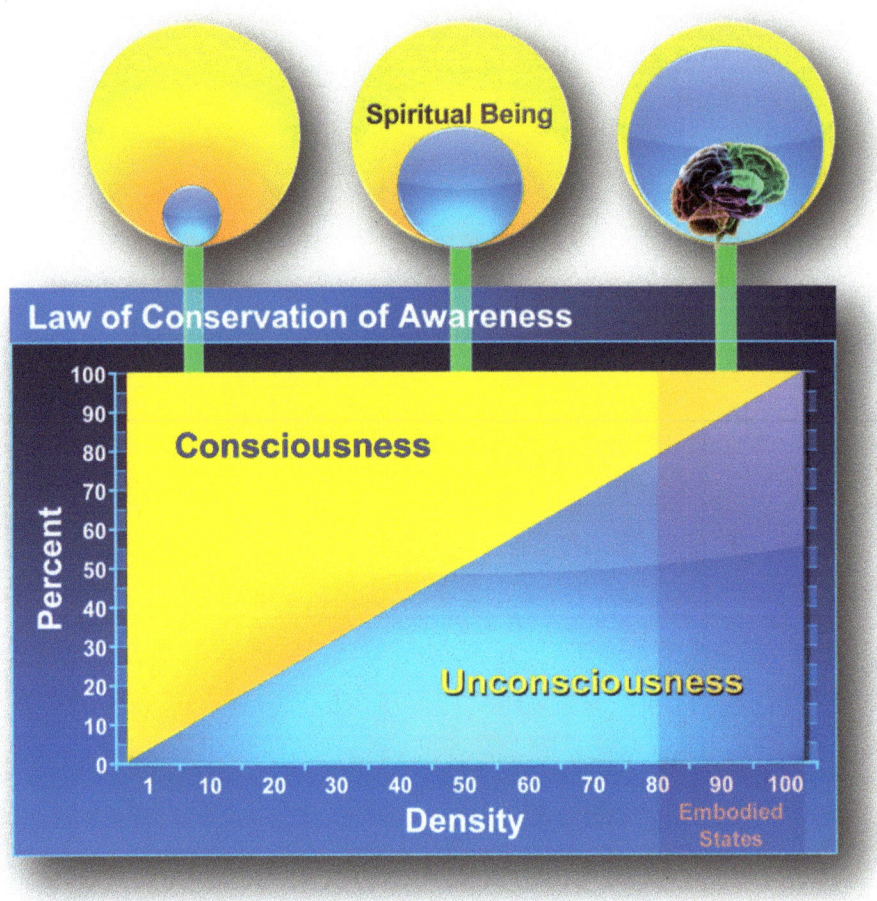

– Fig. 2 –

The diagram in Fig. 2 maps a conscious being's awareness configuration at three points along the density scale.

As stated in Axiom 2, as consciousness condenses, it changes into unconsciousness. As density rises, unconsciousness increases as a proportion of the being's awareness. Unconsciousness and density are for all practical purposes equivalent.

Although almost any level of consciousness can be achieved during an incarnation, the embodied states are usually the densest. In these states, the brain floats within the conscious being's field of awareness, which is largely unconscious.

Cognitive Engine

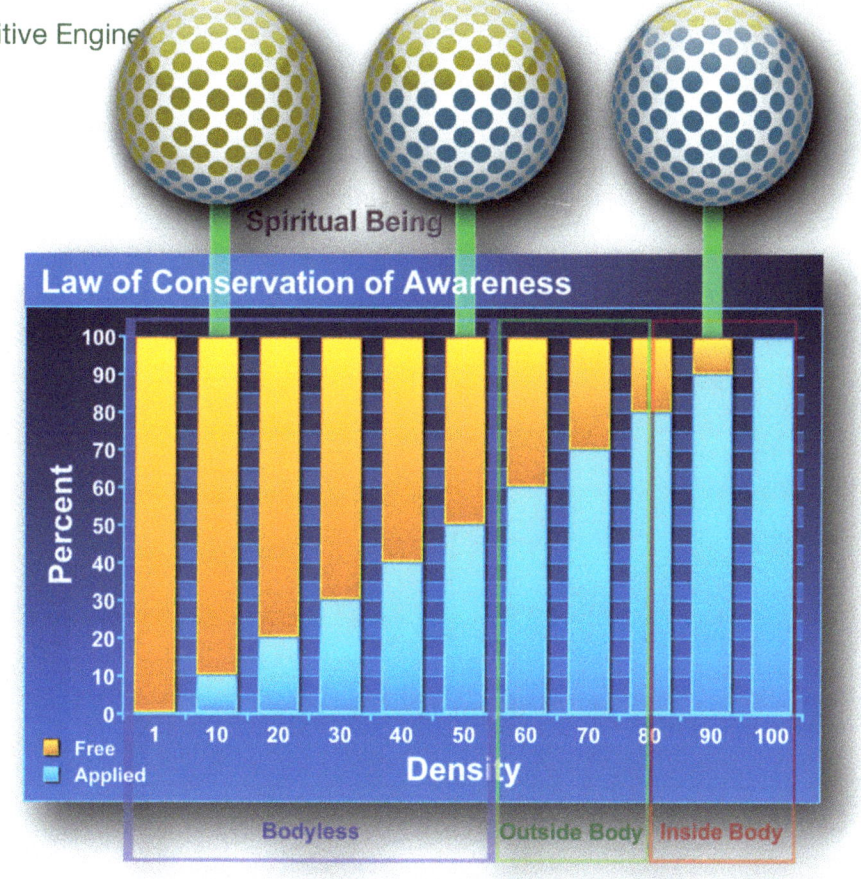

- Fig. 3 –

The diagram in Fig. 3 depicts the three states of being shown on the last page, this time illustrating the density values numerically using dots. The yellow dots represent awareness in a conscious state, and the blue dots represent it in an unconscious state.

As illustrated in Fig. 3, each of the three examples in the diagram represents a major state of being.

Inside Body: The being is animating a body and is occupying the same physical space as the body.

Outside Body: The being is outside the body but continues to animate it. The being feels space surrounding them instead of being inside a solid body. [1]

Bodyless: Although beings can achieve almost any state while incarnated, the top states are usually found in the angelic realms.

---

[1] Many kinds of human out-of-body states are possible. This text will use out-of-body states which occur during meditation as a reference point. Typically, these states arise spontaneously as clean, clear, crisp, exhilarating experiences of awareness, lasting anywhere from seconds to a few hours.

Axiom 6 – Nominal Value
Awareness elements are assigned a nominal value.

For discussion purposes, this text will assume a conscious being in its original state is composed of 100 trillion points of awareness. (You can visualize them as points of light.) This number will be called its *nominal value*, and represented by the symbol n. Assigning a nominal value is convenient, because it allows us to write a formula for the law of conservation.

Formula 1
Law of Conservation of Awareness

$$C = (n - U)$$

Where:
- C = Conscious awareness
- U = Unconsciousness
- n = the number of awareness points per being; nominally 100 terapoints (100 trillion points)

Example 1:
If conscious awareness comprises 20 terapoints, unconscious awareness is 80 terapoints; i.e., 20 = (100 – 80).

Example 2:
If conscious awareness measures 60 terapoints, unconscious awareness is 40 terapoints; i.e., 60 = (100 – 40).

Gold = conscious awareness
Blue = unconscious awareness

When graphed, this formula produces the chart in Figure 1. Awareness changes state from one form to another, but the total amount of awareness in the being does not change, indicating a law of conservation.

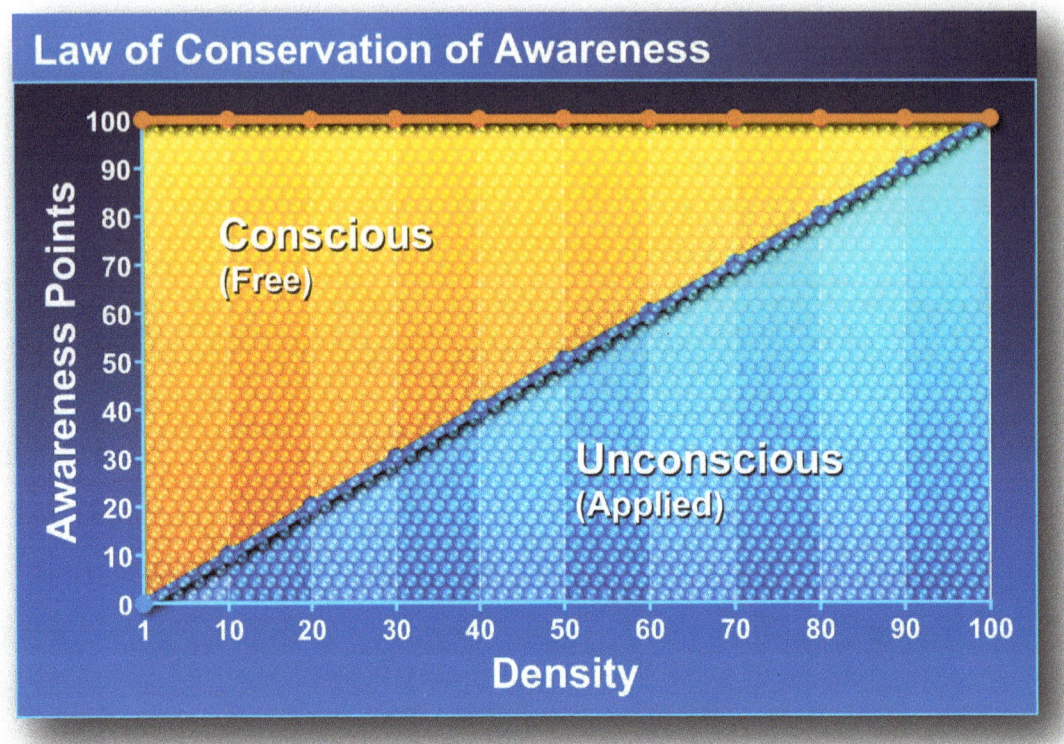

– Fig. 4 –

Now that nominal value has been defined, we can draw a chart like this to illustrate the Law of Conservation. The gold dots represent free awareness points and the blue dots represent applied awareness points. The number of points in each vertical bar remains constant. Only the state of the points changes.

Cognitive Engineering

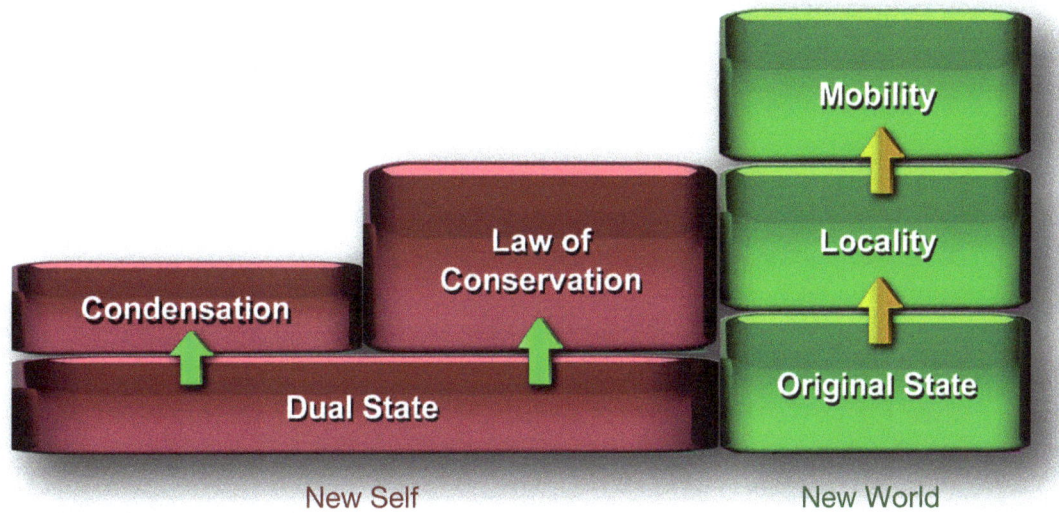

● Summary

The diagram above shows the relationship and dependencies between the six key laws outlined thus far in this chapter. These laws refine our understanding of the *self* and the *world*. To recap:

New Self
- *Dual State* – The two states of awareness are free (conscious) and applied (unconscious).
- *Condensation* – The universe is a condensed form of consciousness.
- *Law of Conservation* – Awareness cannot be created or destroyed, only changed from one form to another.

New World
- *Original State* – In its original state, a conscious being has no location in space.
- *Locality* – A conscious being occupies a location in space through incarnation.
- *Mobility* – Physical distance is not a factor in the interlife.

Cognitive Engineering

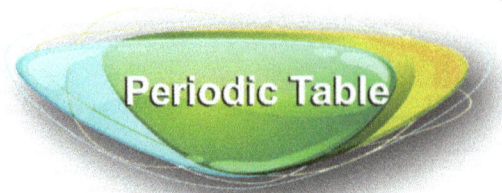

Periodic Table

● Discerning Structure

The convergence of science and consciousness appears to be a global megatrend, and it is likely that humanity will undertake large scale research of consciousness as a scientific discipline in the 21st Century.

During the 20th Century, over a million man years of research went into understanding chemistry and its applications in industry, biology, agriculture and geology. When humanity has made a comparable investment into understanding consciousness, it will have the beginnings of a real science.

Most educated people are familiar with the periodic table of atomic elements, which is the foundation of modern chemistry. The table begins with light elements like hydrogen and helium at the top, and proceeds downward through progressively heavier elements, such as gold and uranium, as atomic weight increases.

Human bodies are made of chemical elements, and human beings are made of awareness elements. A rudimentary periodic table of awareness is presented in this chapter as a starting point for developing our understanding of human consciousness potential.

The first periodic table, made by Lavoisier in 1789, had only 33 elements, and contained errors and omissions. The table in this book has only 22 elements, and contains errors and omissions. Like Lavoisier's first atomic table, our table is of interest because it is a small first step on the correct path. The table illuminates a road that leads to a science of awareness which can yield substantial benefits for humankind. Hopefully, future generations of awareness researchers will fill in the missing pieces.

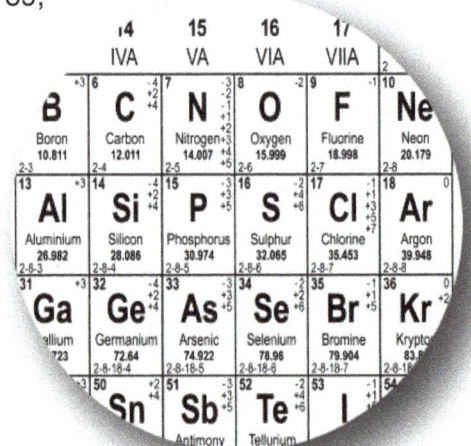

Like the atomic table, the awareness table organizes elements according to their weight and density. It starts at the top with the lightest, most expanded states of being, and proceeds downward through increasing levels of density to unconsciousness.

Cognitive Engineering

In the chemical table, atomic weight is determined by the number of protons and neutrons in an atom's nucleus. (Atoms with larger nuclei have heavier atomic weight.) In the awareness table, protons and neutrons are replaced with two kinds of awareness states: consciousness and unconsciousness. Awareness weight and density are determined by the proportion of unconsciousness in the being. The more unconsciousness a being has, the greater is its density and weight.

Like an atomic nucleus, a conscious being is made up of different combinations of these two fundamental states. Unlike a nucleus, a being can change its awareness configuration, yielding complete mobility across the entire spectrum of states.

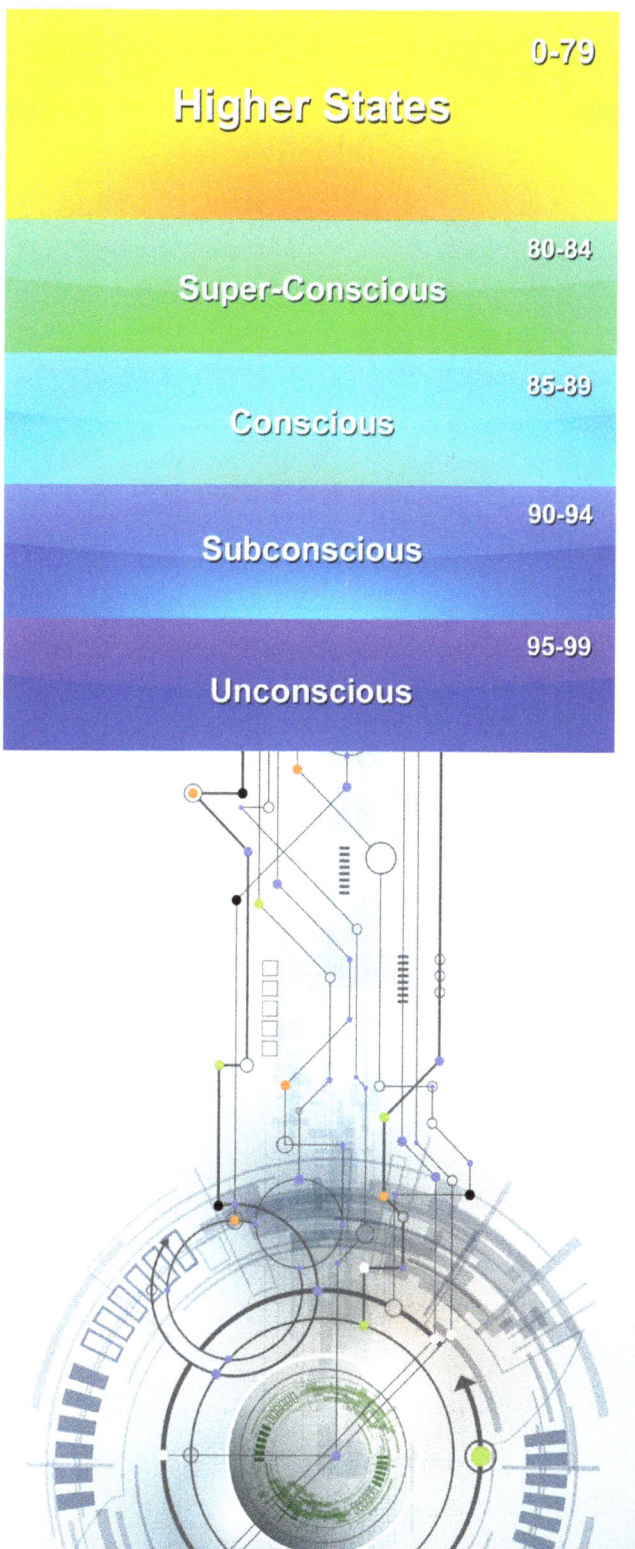

Cognitive Engineering

This chapter illustrates a conscious being's configurations for 20 common states of awareness. A sample illustration is shown here. The sphere represents a conscious being. The dots show its awareness configuration.

There are 100 dots in the sphere. Each dot represents a terapoint. Yellow dots symbolize awareness in a conscious state, and blue dots represent it in an unconscious state.

For the sake of simplicity, this text will assume a conscious being's spectrum of awareness contains 100 unique states numbered 0 to 99. The state number indicates the number of blue dots in a being's composition. This number is assigned the symbol **D** for density. The value of **D** indicates the density of the being, like atomic numbers represent atomic weights in chemistry.

The unconscious is primarily memory, so we can think of the blue dots as memory storage areas. Each memory has a unique energy pattern, like a voice print. The English language would be stored in one of the blue dots. When spoken, each of its million words has a distinct wave signature.

If your "English language" dot is blue, you have the memories to understand the language. It the dot is yellow, you don't. (At the highest states, everything goes. You can't move, think, speak, or read. If you try to read, you only see squiggles.)

Other blue dots contain the memory patterns for opening a door, riding a bicycle, the smell of jacaranda blossoms, and so on.

Cognitive Engineering

Higher States   0-79   Periodic Table Rows 1-16

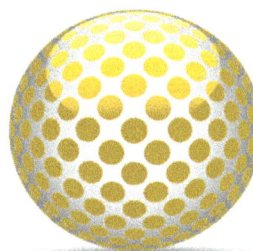

0. Original State
This globe of terapoints represents a conscious being's geography. In our original state, conscious beings are pure awareness.

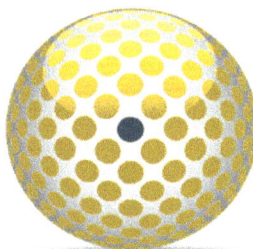

1. Identity
When an identity is first formed, the being experiences their existence as distinct from the universal field of consciousness from which they arose.

States 2 through 79
Above level 79, a being will expand out of their body, and spill over into the surrounding space. At lighter states, the being is completely outside the body, and in the top states, bodyless.

As very high out-of-body states of being are relatively uncommon on Earth, they are not covered in this book.
Human beings usually occupy denser levels of awareness.

Super-Conscious States   80-84   Periodic Table Row 17

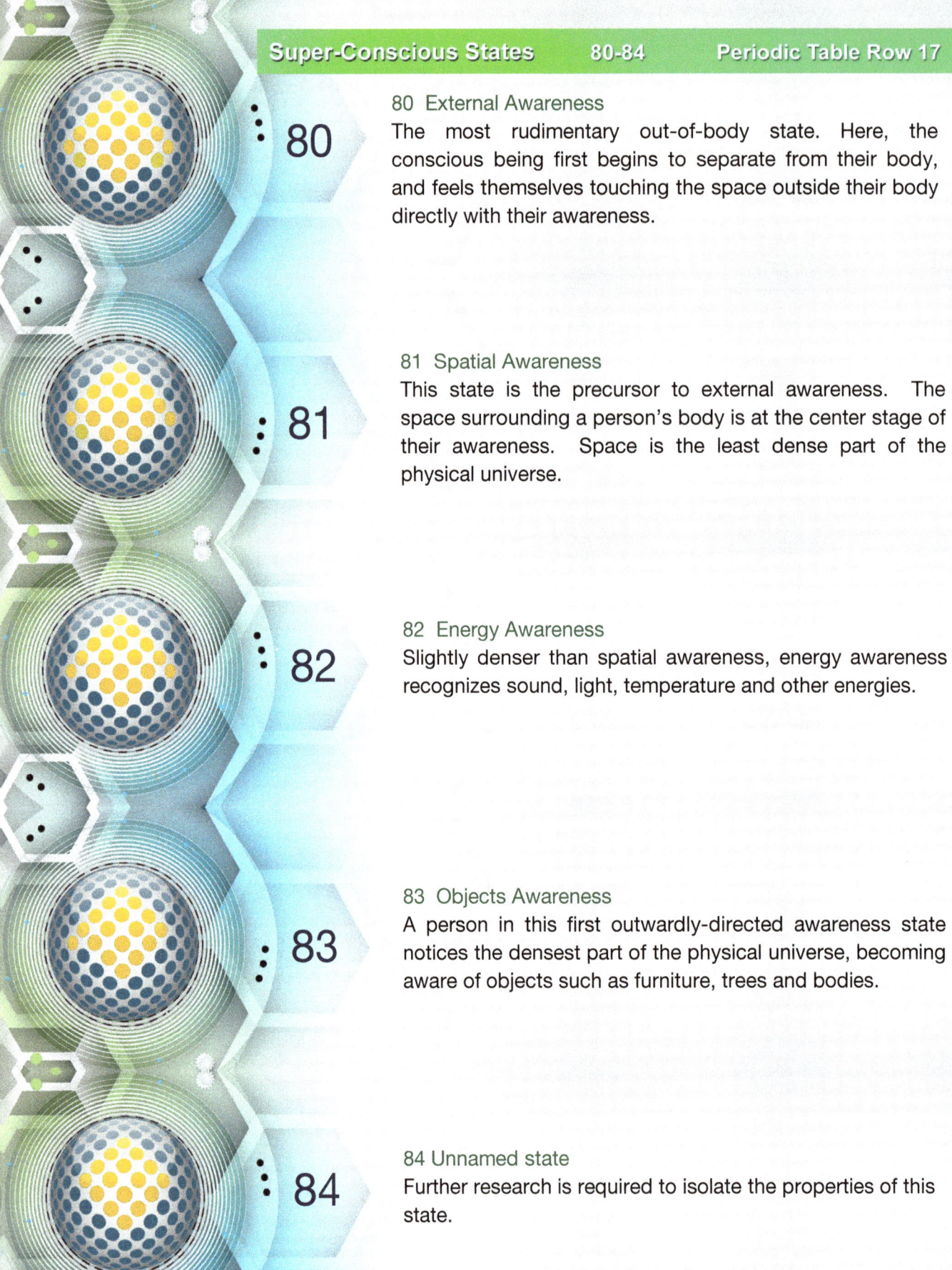

80 External Awareness
The most rudimentary out-of-body state. Here, the conscious being first begins to separate from their body, and feels themselves touching the space outside their body directly with their awareness.

81 Spatial Awareness
This state is the precursor to external awareness. The space surrounding a person's body is at the center stage of their awareness. Space is the least dense part of the physical universe.

82 Energy Awareness
Slightly denser than spatial awareness, energy awareness recognizes sound, light, temperature and other energies.

83 Objects Awareness
A person in this first outwardly-directed awareness state notices the densest part of the physical universe, becoming aware of objects such as furniture, trees and bodies.

84 Unnamed state
Further research is required to isolate the properties of this state.

Conscious States 85-89 — Periodic Table Row 18

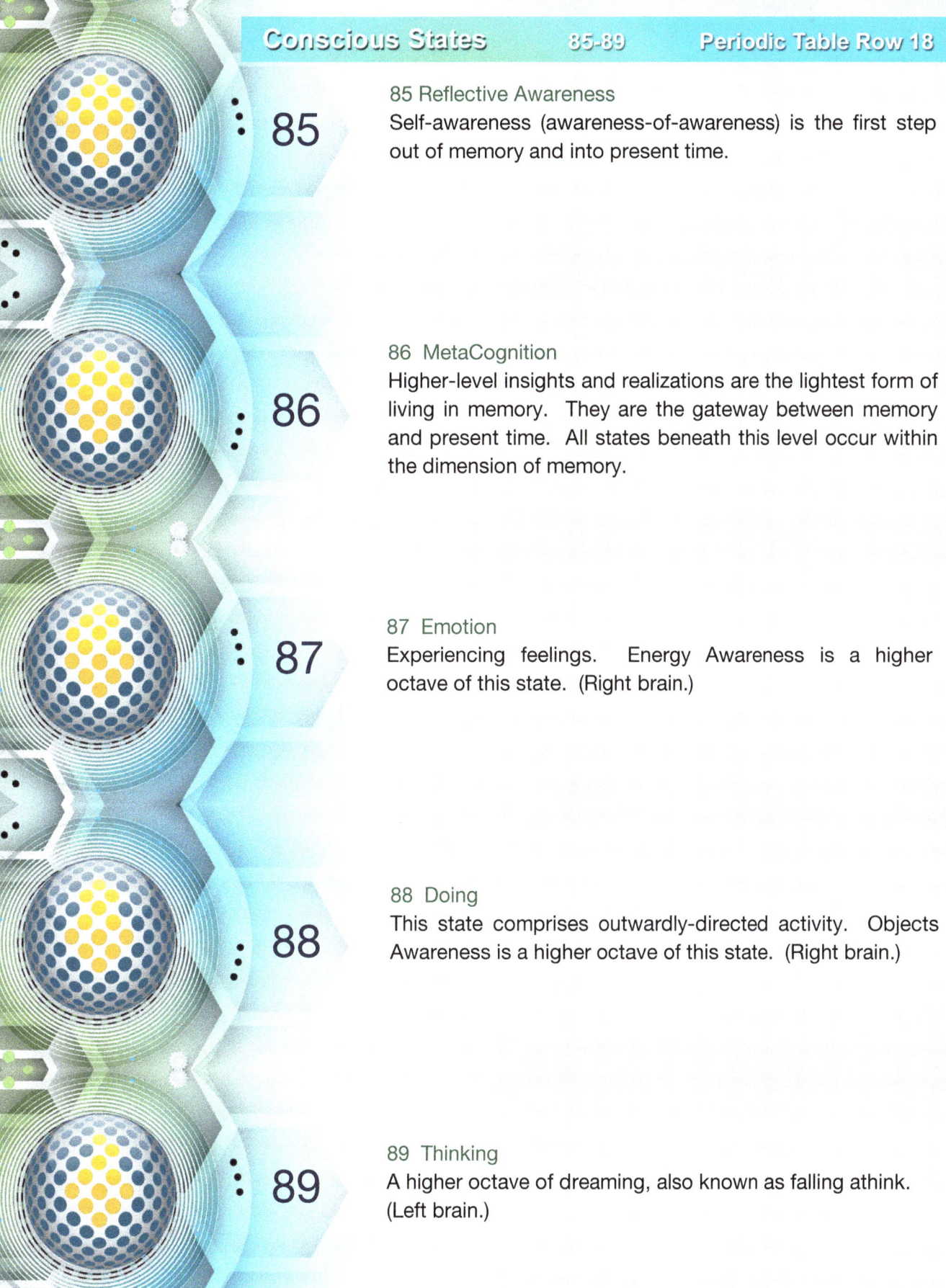

85

85 Reflective Awareness
Self-awareness (awareness-of-awareness) is the first step out of memory and into present time.

86

86 MetaCognition
Higher-level insights and realizations are the lightest form of living in memory. They are the gateway between memory and present time. All states beneath this level occur within the dimension of memory.

87

87 Emotion
Experiencing feelings. Energy Awareness is a higher octave of this state. (Right brain.)

88

88 Doing
This state comprises outwardly-directed activity. Objects Awareness is a higher octave of this state. (Right brain.)

89

89 Thinking
A higher octave of dreaming, also known as falling athink. (Left brain.)

Subconscious States **90-94** **Periodic Table Row 19**

90 Unnamed state
Further research is required to isolate the properties of this state.

91 Unnamed state
Further research is required to isolate the properties of this state.

92 Daydreaming
A semi-lucid state between thinking and light sleep.

93 Light Sleep
Peripheral awareness of self and environment.

94 Dreaming
The lowest form of memorable cognitive activity

Unconscious States   95-99   Periodic Table Row 20

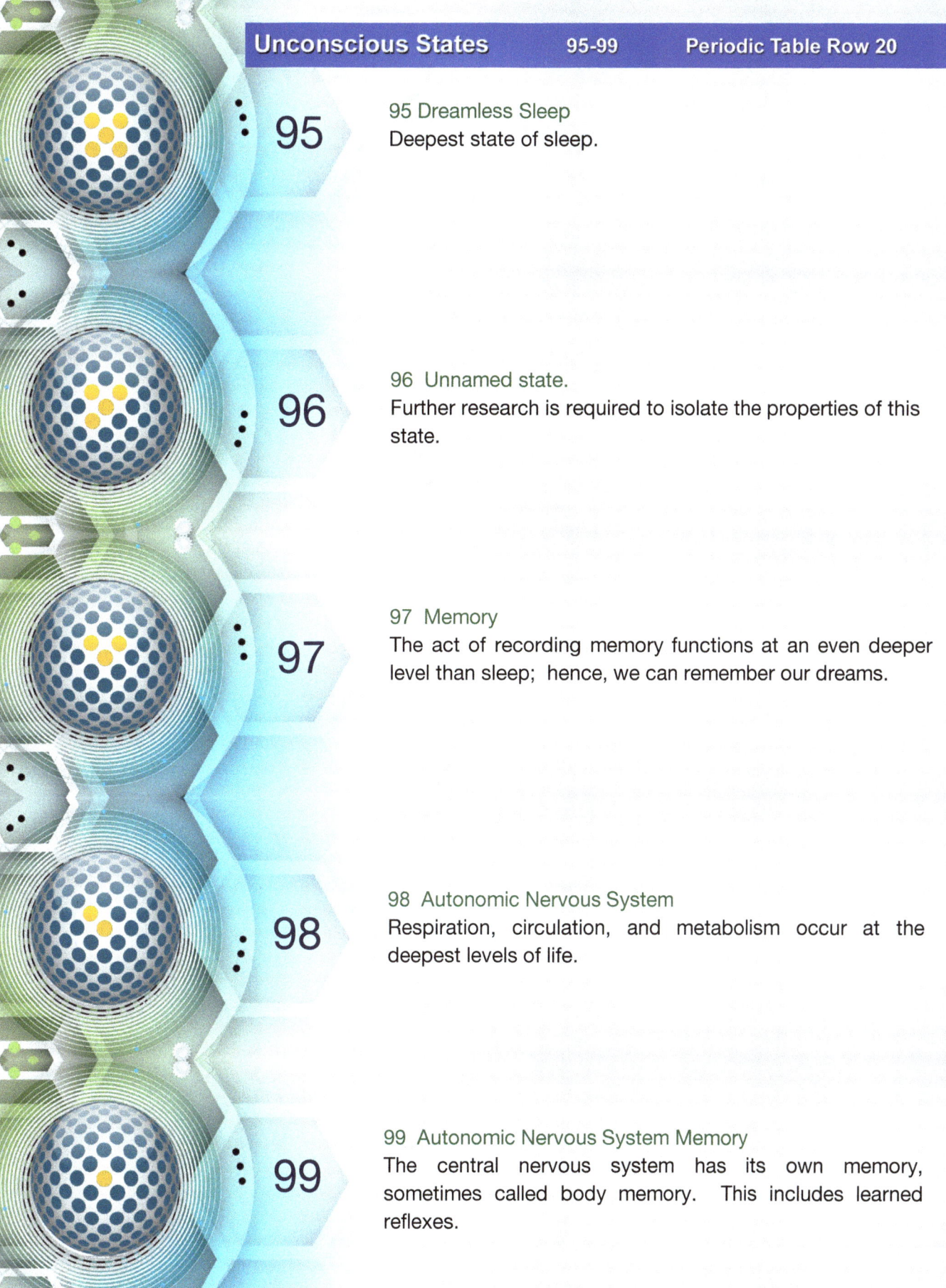

95   95 Dreamless Sleep
Deepest state of sleep.

96   96 Unnamed state.
Further research is required to isolate the properties of this state.

97   97 Memory
The act of recording memory functions at an even deeper level than sleep; hence, we can remember our dreams.

98   98 Autonomic Nervous System
Respiration, circulation, and metabolism occur at the deepest levels of life.

99   99 Autonomic Nervous System Memory
The central nervous system has its own memory, sometimes called body memory. This includes learned reflexes.

Periodic Table of Awareness

Row	A	B	C	D	E	Level
1	O⁰ Original State	I¹ Identity	2	3	4	Higher States
2 to 16	Outside Body States 5 to 79					
17	Ex⁸⁰ External Awareness	S⁸¹ Spatial Awareness	EA⁸² Energy Awareness	OA⁸³ Objects Awareness	84	Super-Conscious States
18	RA⁸⁵ Reflective Awareness	C⁸⁶ Cognition	E⁸⁷ Emotion	D⁸⁸ Doing	T⁸⁹ Thinking	Conscious States
19	90	91	Dy⁹² Daydreaming	SL⁹³ Light Sleep	Dm⁹⁴ Dreaming	Subconscious States
20	Ds⁹⁵ Dreamless Sleep	96	M⁹⁷ Memory	A⁹⁸ Autonomic Nervous System	Am⁹⁹ Autonomic Memory	Unconscious States
	Being	Space	Energy	Matter	Mind	

Higher States — Rows 1-16
- Out-of-body, bodyless and angelic states.
- Pure awareness of awareness.
- Just "I am," not "I am [something]."
- Comprise the majority of states of being.

Super-Conscious States — Row 17
- States 84-81 are awareness of awareness plus [something].
- External Awareness 80 is the doorway to higher consciousness. All Higher States are external awareness states.

Conscious States — Row 18
- States 89-86 are conscious awareness of [something].
- Reflective Awareness 85 is a pivotal threshold state; awakening into present time and awareness of awareness. All states north of this build on it.

Subconscious States — Row 19
- Unconscious awareness of [something].
- Unconscious awareness plus perception (e.g., dreaming).
- "I am [dreaming]" is a denser state than "I am [thinking]."

Unconscious States — Row 20
- Unconscious awareness
- No perception
- Example: Dreamless sleep.

Cognitive Engineering

● Research Directions

Lavoisier determined the sequence of elements in his periodic table by measuring their weights on scales in his laboratory. Comparable instruments do not exist today for measuring the properties of awareness, so the numbers assigned to elements in our table represent hypothetical examples. The preceding illustration shows an example of how a periodic table might look in the future when measurements can be made. Capable instruments will inevitably arise from the inexorable progress of microelectronics driven by Moore's Law. When this occurs, the periodic table in this book can be updated to reflect empirical data.

● Periodic Table Rows

Science can and should define a hundred states of being arranged by order of density into an organized system as shown by the table on the previous page. The specific numbers assigned to the states in this table may be speculative, but the underlying architecture of arranging states by density is sound.

The table on the preceding page summarizes the characteristics of the periodic table rows. (Read it from the bottom up.) Awareness is cumulative. Each level builds upon the previous one. For example, when you open a door in Doing, you are being [opening a door]. When you open a door in Objects Awareness, you are being [y o u], incidentally opening a door. (In this book, "you" means "your awareness," not a personality.)

Objects Awareness 83 is north of Reflective Awareness 85. It contains and builds on the awareness of awareness established in Reflective Awareness. A person opening a door in Objects Awareness is aware of their existence as awareness in present time. The focus has shifted from being the action to being awareness.

This principle of awareness accumulation applies across the entire table. For example, all states north of External Awareness 80 are out of body states and contain external awareness of awareness. Each successive state adds more awareness to the previous one.

Cognitive Engineering

● Periodic Table Columns

If matter exhibits periodicity, and the physical universe is a condensed form of awareness (per axiom 2), then awareness itself may also exhibit periodicity. The laws of matter may be shadows of the laws of awareness.

The table's columns as shown are arranged in five elemental groups: being, space, energy, matter and mind (or spirit, air, fire, earth and water). As we examine the states in each column, we notice some interesting evidence of periodicity. Each state of being may have an affinity to one of the five elemental groups, and these patterns of affinity may repeat. For example:

A. Being: Original State, External Awareness, Reflective Awareness and Dreamless Sleep are all awareness states relating to beingness.

B. Space: Spatial Awareness obviously involves physical space while Identity may be thought of as the container of beingness volume.

C. Energy: Emotion is an energy state and memory recordings are energy signatures.

D. Matter: Doing normally involves interaction with material objects, while Objects Awareness involves awareness of them.

E. Mind: Thinking and Dreaming are awareness states relating to mental activity (one being a lighter form of the other). Autonomic Memory could be thought of as the body's mind.

In examining the periodic table, we also find there are clearly some states that seem to be higher octaves of others; for instance, thinking and dreaming, energy awareness and emotion, objects awareness and doing, and external and reflective awareness.

Both of these observations – affinity groups and higher octaves – reveal characteristics of periodicity. Future researchers will provide empirical data which enables the construction of a conclusive table.

Cognitive Engineering

Density and Volume

● Overview

How does a being's density affect its volume? The diagram on the facing page shows the answer. It illustrates beingness volume plotted against an awareness scale from 1 to 100.

As shown by the green arrow on the left, the volume of a conscious being expands in proportion to its awareness level. The volume awareness occupies is directly proportional to the amount of awareness. Beings in higher states expand to fill greater volumes of awareness.

Conversely, a conscious being's volume diminishes as its density increases, as indicated by the yellow arrow on the right of the diagram. Density of awareness is inversely proportional to the amount of volume the awareness occupies. Less aware states of consciousness occupy less volume and experience denser states of being (for example, incarnation).

The size of a being's awareness volume is inversely proportional to its density. Someone at density level 90 has more density and less volume than someone at density level 50. Density increases as volume diminishes.

Density is an attribute of unconsciousness. The amount of consciousness (**C**) is the reciprocal of unconsciousness (**U**). As mentioned earlier, unconsciousness (**U**) and density (**D**) are equivalent. Given Formula 1, then, one would expect density to grow as unconsciousness increases.

$$C = (n - U)$$

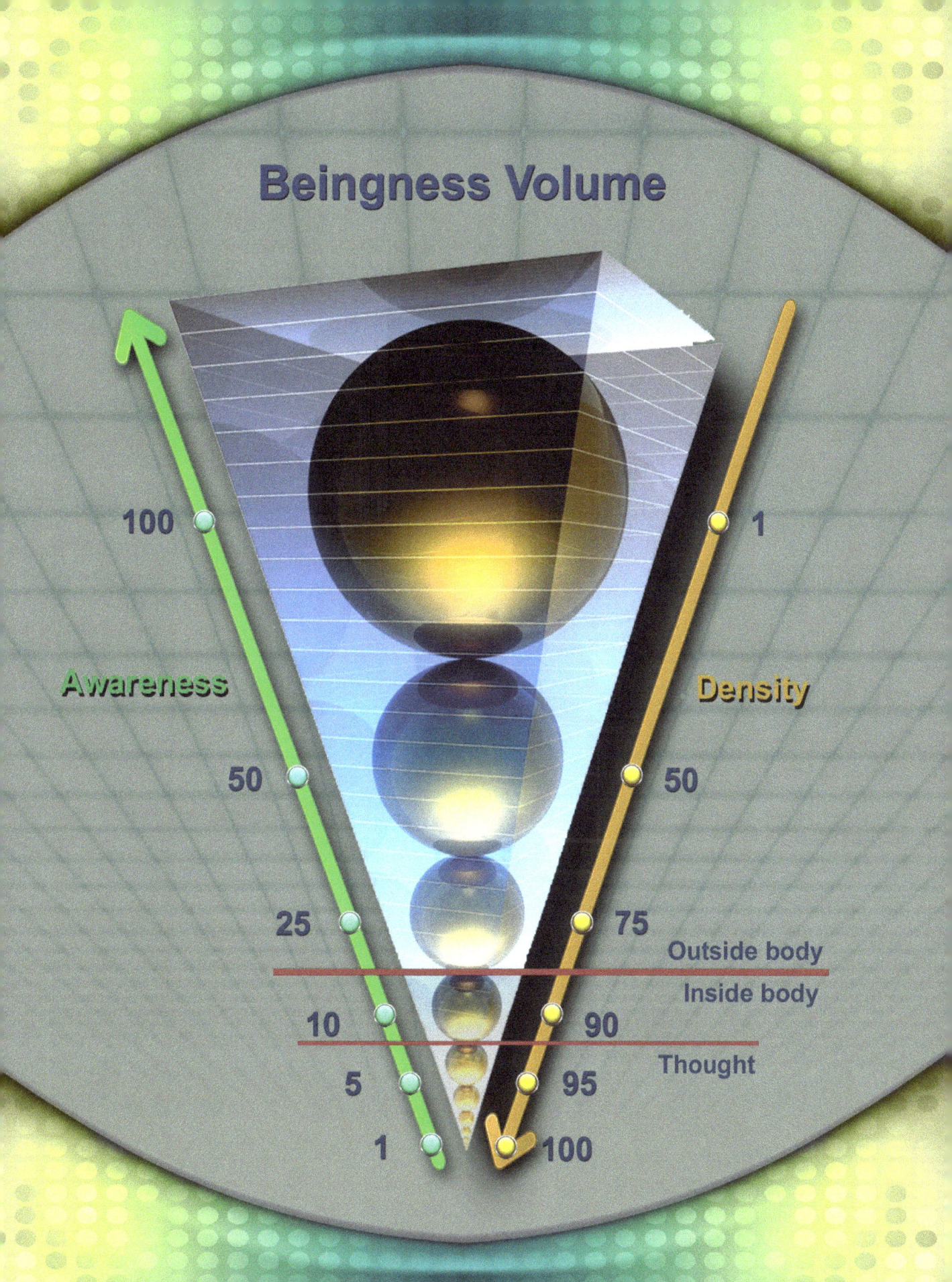

BEINGNESS VOLUME

This diagram illustrates 3 examples of expanded states of being. Many other different kinds of experiences are possible.

The golden globe symbolizes the spiritual being.

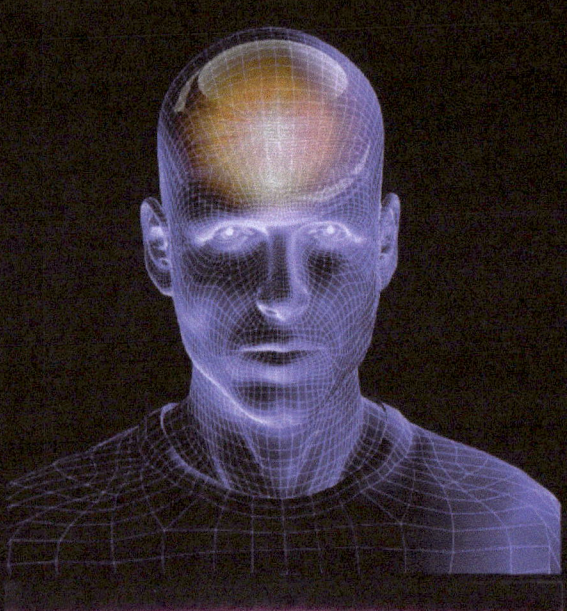

THINKING

10 / 90 — "Normal" human awareness level. The being is completely inside their body.
Conscious: 10%
Unconscious: 90%

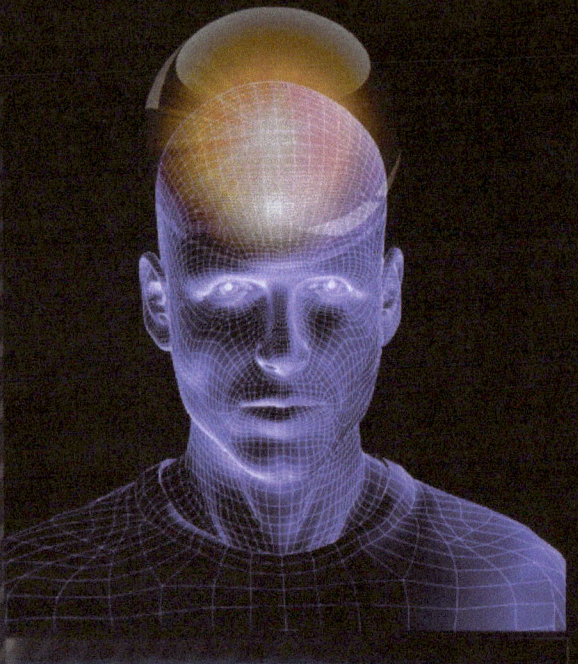

EXTERNAL AWARENESS

20 / 80 — Initial contact between the being and the space surrounding their body. First step into external states.
Conscious: 20%
Unconscious: 80%

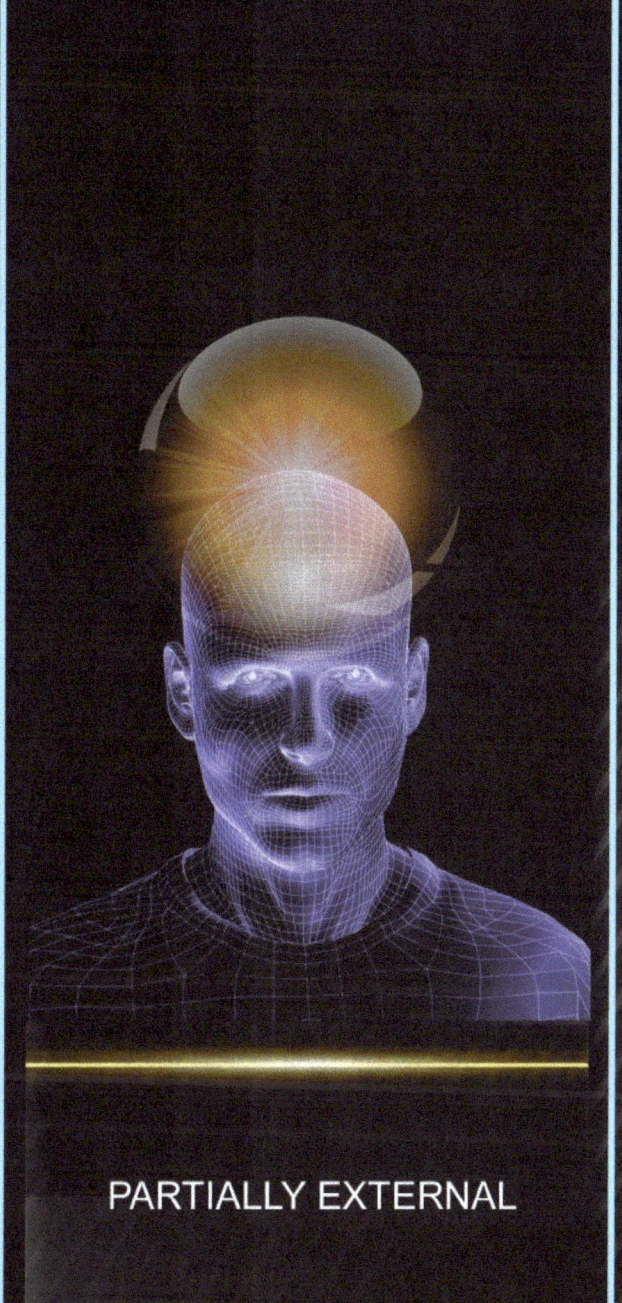

PARTIALLY EXTERNAL

30
70

As awareness grows, the being's volume continues to expand beyond the space their body occupies.
Conscious: 30%
Unconscious: 70%

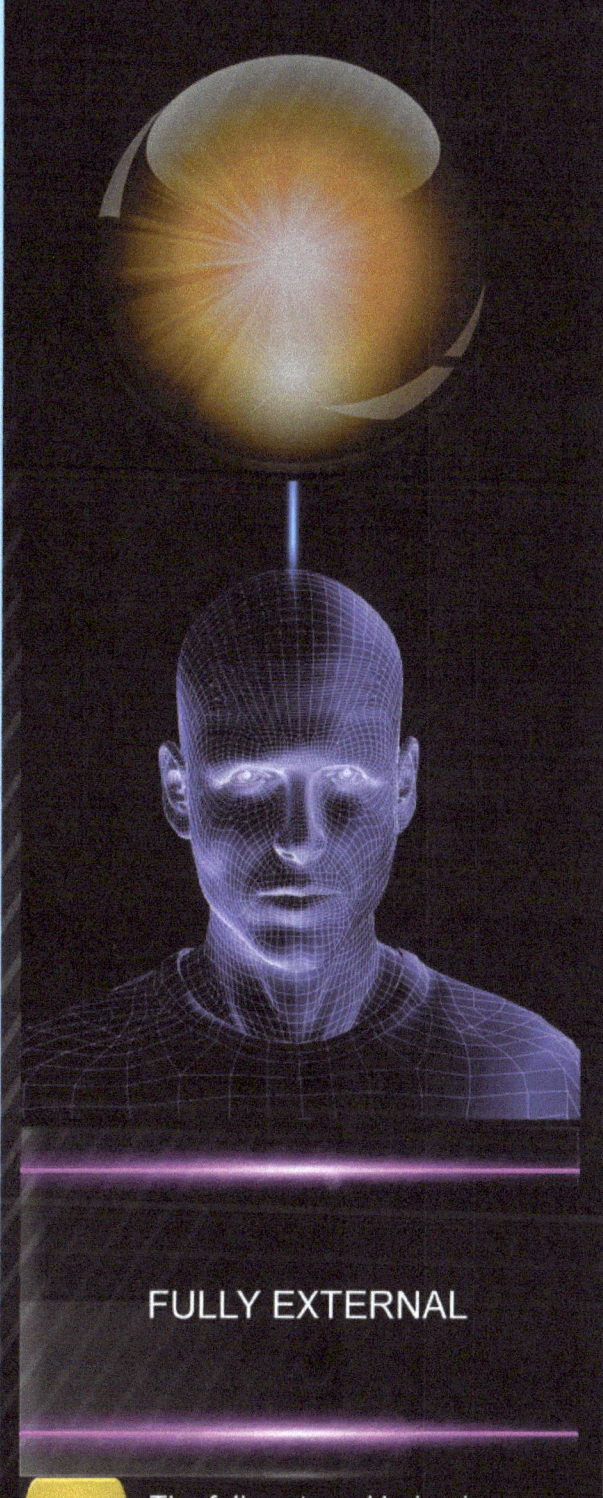

FULLY EXTERNAL

40
60

The fully external being is connected to its body by thin silver cord. Although the cord shown here is short, it can actually be several feet long. Sixty percent of states are above this level.
Conscious: 40%
Unconscious: 60%

Cognitive Engineering

In Figure 3 in the Law of Conservation chapter, three sample states of awareness were illustrated as shown below for simplicity's sake.

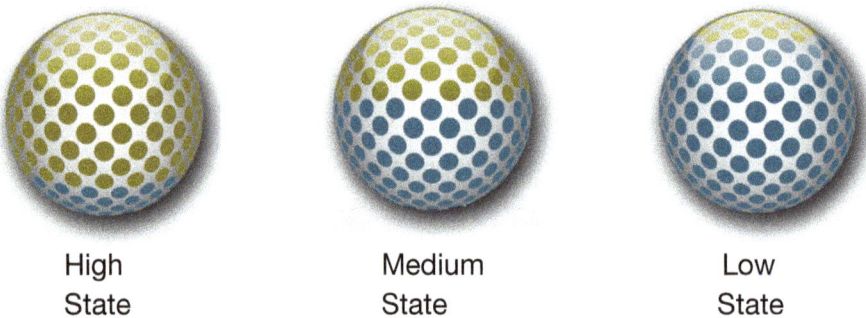

High State | Medium State | Low State

Actually, beingness volume is proportional to awareness level, like this:

High State | Medium State | Low State

If we say the radius of the sphere r is approximately equal to the conscious awareness level of the being C, then we can write a formula for volume:

Cognitive Engineering

Formula 2
Beingness Volume

$$BV = \frac{4}{3}\pi r^3$$ Where: BV = Beingness Volume
  r = Radius ($=C$)

This formula allows us to approximate the volume dimensions of various states of being. It is simply the formula for volume of a sphere, developed by the ancient Greek philosopher Archimedes.

The beingness volume formula is not actually measuring the volume of awareness, because technically, a being has no dimensions. It is measuring the volume of space which the being displaces. From this measurement, we can infer the dimensions of the being.

The being is not part of the field of space which surrounds it. It is an ambassador from the awareness universe above. No part of the being is in space. The being is a place where space is not. It is a hole in space filled with awareness. Space forms itself around it.

● Examples

Following are two examples which illustrate how to calculate beingness volume for two different values of r.

Example 1: $r = 25$

Constants: 4/3 = 1.33 and π = 3.14
1.33 X 3.14 = 4.18

Variables: $r = 25$
      $r^3 = 15,625$

Formula: BV = 4.18 X 15,625

Answer: BV = 65,312

Example 2: $r = 75$

Constants: 4/3 = 1.33 and π = 3.14
1.33 X 3.14 = 4.18

Variables: $r = 75$
      $r^3 = 421,875$

Formula: BV = 4.18 X 421,875

Answer: BV = 1,763,437

● Conclusions

The facing diagram summarizes the relationship between the Law of Conservation of Awareness, Beingness Volume, and the Periodic Table of Awareness.

As shown, beingness volume expands proportionally with conscious awareness. Larger volumes have less density. A conscious being's volume expands in proportion to its level of awareness, and contracts in proportion to its degree of density.

The periodic table in this book describes condensed states of being which currently prevail on Earth. Unconscious awareness is the overarching component of these states.

When viewed from higher states of being, human existence is unimaginably dense and unaware. True life as a conscious being only begins outside the body. Approximately eighty percent of the awareness scale is out-of-body states.

Higher states of being are attainable which offer expanded conscious awareness and greater volumes of awareness. Although relatively few people in human history have reached the top states, anyone reading this book is likely to have the capacity to go all the way.

We know that awareness is scalable; the age-old question is how to make it sustainable. This book presents a radical new set of strategies for achieving permanent gains.

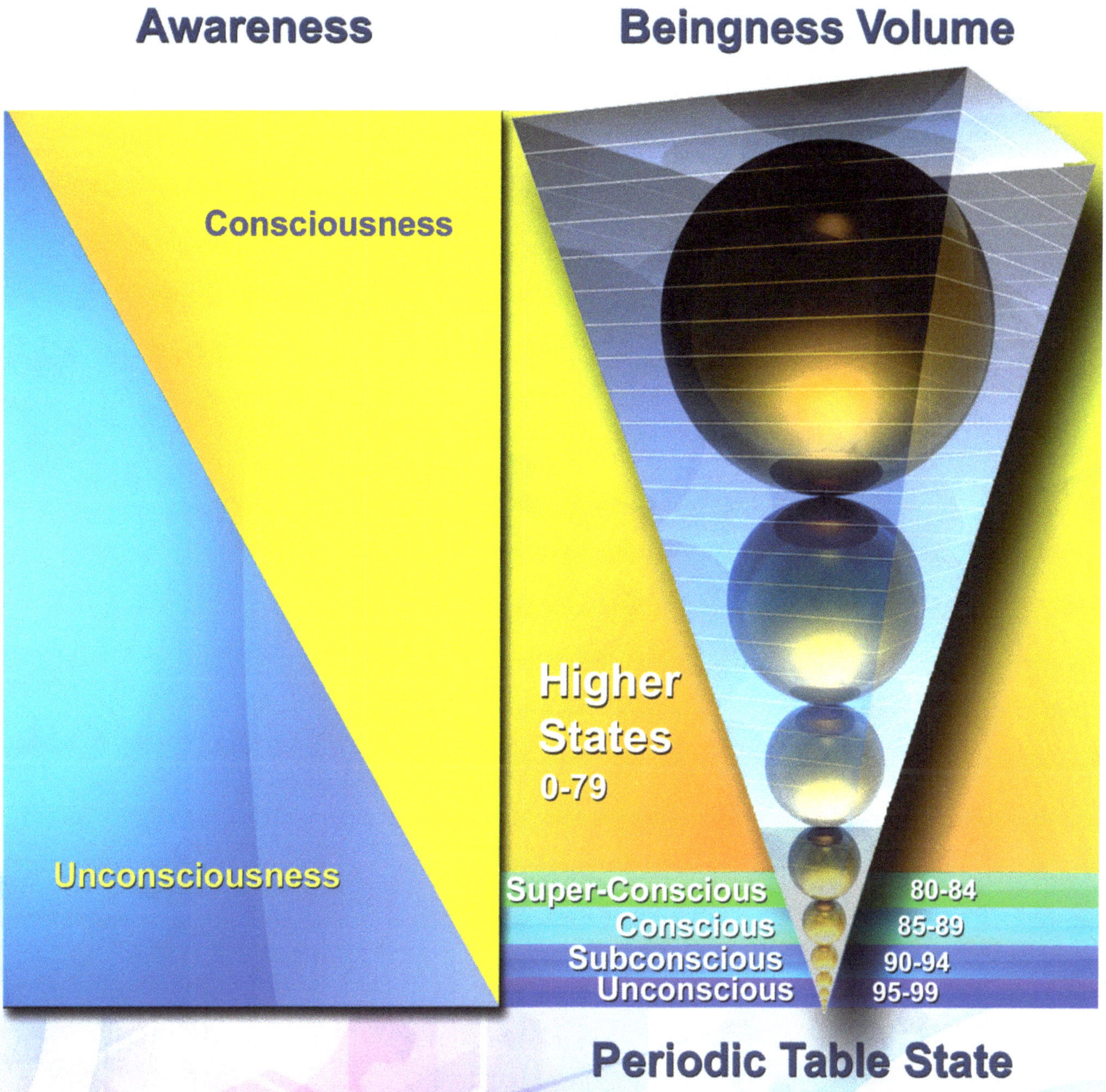

Awareness   **Beingness Volume**

Consciousness

Higher States
0-79

Unconsciousness

Super-Conscious   80-84
Conscious   85-89
Subconscious   90-94
Unconscious   95-99

Periodic Table State

Memory is the dimensional interface

between consciousness and the physical universe.

3

Macro Memory

Introduction

● **Orientation**

To restore a conscious being's awareness from an unconscious state to a conscious state, it is essential to understand the nature and composition of unconsciousness. Per Axiom 2, unconsciousness is primarily expressed through memory. Hence, we need to understand the mechanics of a conscious being's memory in order to treat unconsciousness.

● **New Foundation**

Macro memory is defined as a conscious being's capacity for carrying non-local memories with them over multiple lifetimes.

In conventional, personality-based psychology, it is thought to be impossible for someone to remember anything beyond one lifetime. In the traditional view, consciousness and memory are merely functions of the brain. Thus, it is thought that memory cannot extend beyond the boundary of a lifetime.

The growing acceptance in the West of the non-local nature of consciousness is calling the traditional reductionist views increasingly into question. If consciousness exists independently of the brain, then the old models are inaccurate.

If consciousness is separate from the brain, then the conventional view that memory must end with a lifetime is not necessarily true. But if the old theory is incorrect, what is the new one, and how is it validated?

Past life regression therapy has produced a large and compelling body of anecdotal evidence supporting macro memory. However, when it comes to scientific explanations for how macro memory actually works, we are looking into a vacuum. There are none.

The first step in explaining macro memory is to modernize our understanding of the origin of consciousness.

Cognitive Engineering

Per Axiom 0, people are conscious beings inhabiting physical bodies. Except for the body's sensory perceptions, all mental activity in the brain originates with the conscious being. The conscious being takes this mental activity with it when it departs. Trans-life psychology is viable because the mental changes people make today can stay with them beyond their current lifetimes, and continue to influence their conscious and unconscious behavior.

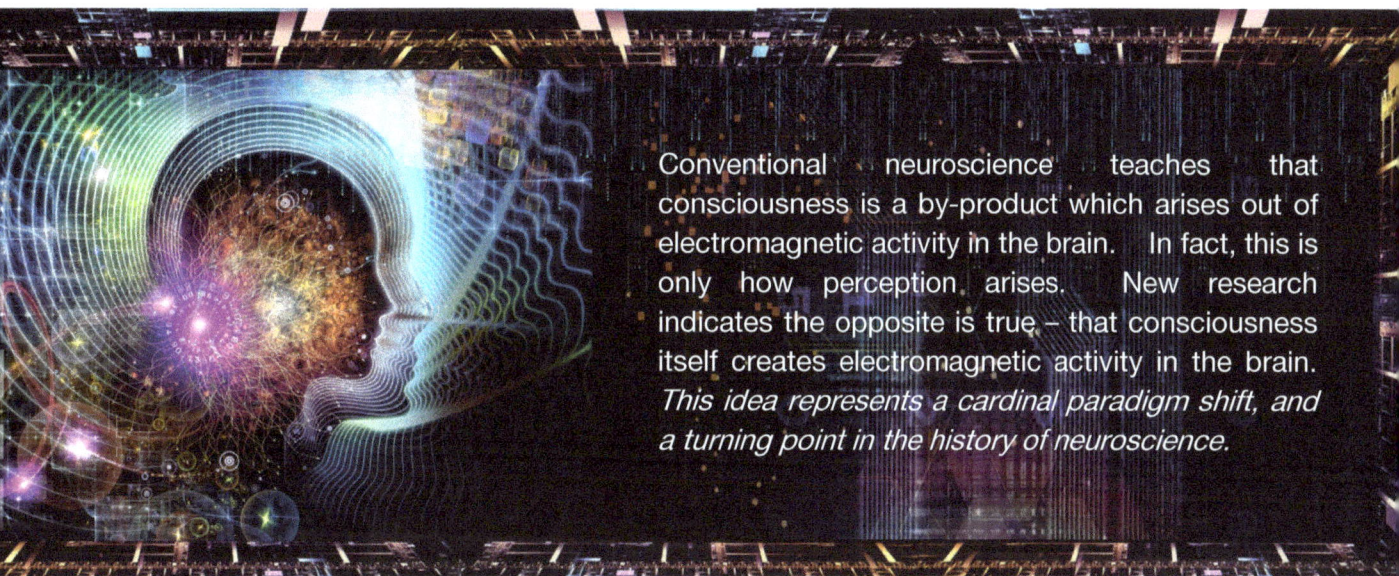

Conventional neuroscience teaches that consciousness is a by-product which arises out of electromagnetic activity in the brain. In fact, this is only how perception arises. New research indicates the opposite is true – that consciousness itself creates electromagnetic activity in the brain. *This idea represents a cardinal paradigm shift, and a turning point in the history of neuroscience.*

Previously, the only memory science available was biochemical neuroscience, which states that memory is part of the brain. In that model, there was no way to separate memory from the brain. Today, pioneering research has built a bridge which allows us to cross between the two worlds of mind and brain.

Cognitive Engineering

● Scope

Memory is the key enabling dimension of embodied life. It is an enormous subject, encompassing all collective knowledge and experience since the beginning of time.

A rigorous treatment of the topic would include not only our personal memories, but also race memory, the collective unconscious, ancestral memory, instincts, body memory, autonomic nervous system memory, cellular memory, and DNA memory, to name a few.

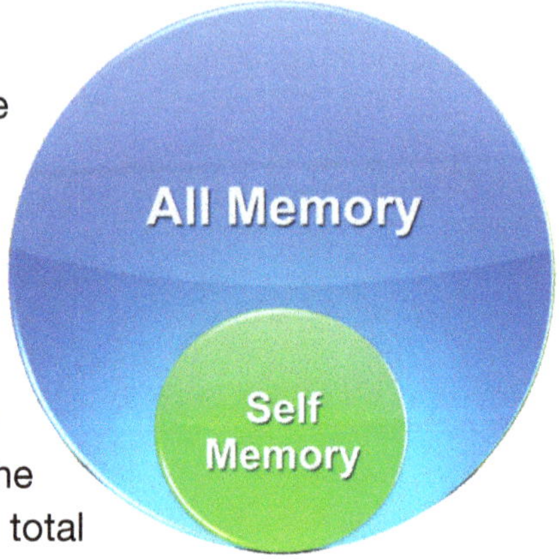

A book like this cannot even begin to address the subject in its entirety, nor will it attempt to. Our scope is strictly limited to practical, applied scientific methods for elevating consciousness.

Accordingly, this text will only consider an individual conscious being's own personal memories, which are labeled *Self Memory* in the diagram, and represent a small subset of the total memory dimension.

This constraint provides a practical framework for therapeutic advances to address macro memory. (The world may be interconnected, but for the most part, your brainwaves are your brainwaves, not somebody else's.)

● Pioneering Breakthrough

Towards the end of the twentieth century, a pioneering experiment at Yale and Stanford paved the way for the effective design of consciousness therapies today.

John Huguenard of Stanford University and David McCormick of Yale University studied how neurons transmit tiny brainwaves at a cellular level. They measured the properties of these waves with scientific instruments, and their findings were published by Oxford University. [7]

The experiment was unique in three ways:

---

[7] *Electrophysiology of the Neuron*, Huguenard and McCormick, Oxford University Press, 1994

1. Although a great deal of research had been done on individual neurons, most of the papers were neurobiology studies of chemical and molecular interactions, not electric fields.

2. The vast body of EEG research studying electrical fields in the previous three decades concentrated primarily on the high-level, global aspects of millions of neurons collectively.[3] In contrast, Huguenard and McCormick zeroed in on the individual neuron.

3. The two men produced the world's first generally-available interactive computer program simulating the neuron's electrical behavior.

Huguenard and McCormick concentrated on the neuron's energy fields, and found that flowing electrons in the brain's neural networks are accompanied by small electromagnetic waves.

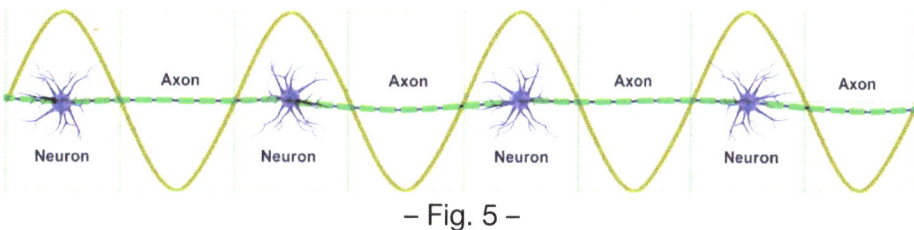

– Fig. 5 –

Fig. 5 illustrates an electromagnetic wave (gold) traveling along a neural pathway made of neurons (blue) and axons (green). Their experiment measured the strength of these neural electromagnetic waves at around 55 millivolts and 5 nanoamperes. (A nanoampere is one billionth of an ampere, so 5 nanoamperes is a very subtle electromagnetic field.)

This important work in neurophysics codified the neuron's electromagnetic characteristics, opening the way for new insights into how human conscious beings interact with the brains we inhabit.

Although valuable, the many earlier studies of individual neuron biochemistry did not provide an entrance point for developing an understanding of how conscious beings interface with our physical neurology. A human conscious being cannot interact directly with solid matter like a molecule, but it certainly can influence

---

[3] Most notably *Electric Fields of the Brain*, Paul L. Nunez, Oxford University Press, 1985

Cognitive Engineering

subtle energy fields. When these tiny neuron-level energy fields are brought into full view, a doorway opens.

This section of the book will explain how the brain's cellular-level electromagnetic fields serve as the interface zone between consciousness and physiology. Moreover, it will demonstrate how the interactions between awareness and matter that occur in this zone provide a scientific basis for macro memory.

A new discipline called *cognitive neurophysics* will be introduced which illuminates the relationship between the conscious being and the neurology it inhabits. Cognitive neurophysics presents a scientific explanation of the mechanics of macro memory. It complements and expands traditional biochemical neuroscience, which is still absolutely essential and valid at the cellular level.

● Dispelling Myths

To avoid confusion as we travel along the road to a new science, it will be helpful to wipe the slate clean of several myths of the past:

1. Intelligence: Traditional neuroscience has taught people to believe the cerebral cortex is the seat of human intelligence. This is incorrect. The conscious being is the seat of human intelligence. All intelligence and life force comes from the conscious being.

2. Learning: A principle of biological neuroscience called Hebbian theory states that learning occurs because repeated stimulation of related neurons raises their synaptic efficiency, or put more simply, "neurons that fire together wire together." Although associated neurons do "wire together," this behavior is an outward manifestation of learning, not a cause.

Learning occurs when the conscious being links related memories together in its mind. When the being's mental field is superimposed on its neurology, the neurons activated by the linked memories will "wire together." Form follows function.

3. Brainwaves: The general public is told that human brainwaves fall into four major categories – delta, theta, alpha and beta – which range from 1 Hertz to 40 Hertz as measured by EEG. There is actually no such thing. If you ask an EEG expert, he will tell you these numbers are mathematical abstractions generated by EEG machines. The raw EEG data actually consists of millions of tiny electromagnetic waves. The EEG machine samples this data and generates averages from the aggregate totals, which it then reports as delta, theta, alpha and beta.

Only the raw EEG data is real. These miniscule electromagnetic waves are generated by the conscious being, either directly, or through its animation of the body's perceptual systems. Science has actually been measuring the electromagnetic wave output of conscious beings for decades without realizing it.

Huguenard and McCormick's research revealed that these tiny waves travel through the brain's neural networks at around 55 miles per hour. This speed is so slow because the brain processes information at electro-chemical speeds. (The brain makes up for its slow transmission speed with its massively- parallel processing architecture.)

Per Axiom 0, a person's true identity is an immortal conscious being, and all other assumptions are false. Conventional, personality-based psychology has proceeded in the absence of the correct starting assumption. Hence, its capacity for delivering value has been limited to addressing short-range issues. Begin with the correct assumptions, however, and startling results emerge on a grander stage.

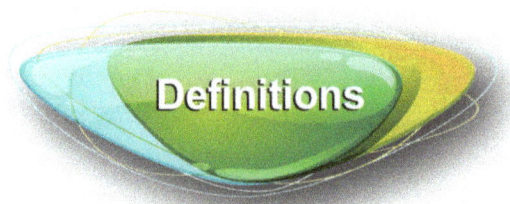

Definitions

🟢 Overview

This chapter introduces the basic terminology used in macro memory science. There are 5 new terms:

        Free awareness point   Active memory particle
        Inert memory particle   Memory wave
        Applied awareness point

The diagram on the facing page illustrates these 5 fundamental elements and their relationships.

Free awareness point: Consciousness is comprised of *free awareness points* which are timeless and formless.

Inert memory particle: Memory is comprised *of inert memory particles* which contain inactive memory recordings.

Applied awareness point: Free awareness points leave conscious awareness and are applied to specific forms, such as memories.

Active memory particle: When an awareness point is applied to an inert memory particle, it changes state into an *active memory particle*, animating the recorded memory.

Memory wave: The active memory particle emits a live *memory wave*.

The following pages explain these basic concepts in more detail.

Trans-Life Memory Science

Conscious

Free awareness point leaves the main body of consciousness...

Free Awareness Point:
Consciousness is comprised of free awareness points which are formless, motionless, unchanging units of pure awareness that exist above time and space.

Applied Awareness Point:
Enters a form such as a memory particle, permeating and animating it.

Unconscious

Inert Memory Particle
Memory waves collapse into inert memory particles when they are recorded. Memory is comprised of inert memory particles which contain inactive memory recordings. Each inert memory particle carries a condensed representation of one memory wave, including its frequency signature and amplitude.

Active Memory Particle
When an awareness point is applied to an inert memory particle, the particle changes state into an active memory particle, and emits a replica of the memory wave which it originally recorded. The applied awareness point is the force of the memory wave's projection.

Memory Wave
A spiritual being's internal representation of experience. It possesses a non-physical frequency and amplitude, without mass, energy or a location in space. Each memory wave has a unique waveform pattern, like a voiceprint.

Pre-Unconscious

FUTURE LIFE INSTITUTE

Cognitive Engineering

1. Awareness Points

An awareness point is a unit of awareness as described in Axiom 6 (restated below).

Axiom 6 – Nominal Value *(restated)*
Awareness elements are assigned a nominal value.

For discussion purposes, this text will assume a conscious being in its original state is composed of 100 trillion points of awareness. (You can visualize them as points of light.) This number will be called its *nominal value*, and represented by the symbol n.

The size of the being's conscious awareness field is proportional to the number of free awareness points enclosed within the being.

2. Awareness Point States

An awareness point has two states – free and applied – as described in Axiom 1 (restated below).

Axiom 1 – Awareness Dual State *(restated)*
Awareness has two states: free and applied.

The two forms of awareness are *free awareness*, which is conscious, and *applied awareness*, which is unconscious.

Free awareness: Free awareness is conscious and formless. It contains no mass, energy, space, wavelength or amplitude. It is stationary, changeless, and exists outside of time and space.

Applied awareness: Applied awareness is unconscious and has a form. It is located in time and changes state in time. Its changes in state over time are called oscillations, and they form waves called memory waves.

Awareness changes state from *free* to *applied* when free awareness is applied to a form, such as a memory. The applied awareness permeates and animates the form. In this process, the free awareness becomes unconscious.

When applied awareness is removed from a form, such as a memory, the awareness reverts back to its original free and conscious state.

Cognitive Engineering

Applied awareness powers the functions of memory and physical life. It could also be called "mind" or "mental energy."

2A. Free Awareness Point

A free awareness point is a formless, unchanging, motionless, stationary unit of pure conscious awareness which exists above time and space.

2B. Applied Awareness Point

An applied awareness point is an awareness point which has left the main body of conscious free awareness points and entered a form, such as a memory wave. The applied awareness point permeates and animates the form.

Memory waves like the one depicted here are powered by the energy of applied awareness points. Memory waves and applied awareness points are unconscious.

3. Memory Wave

A memory wave is the conscious being's internal representation of experience. Memory waves have no mass, energy or location in space, but they do possess a non-physical frequency and amplitude.

Frequency is the universal carrier of information. For example, we recognize color by the frequency of light, and language by its sound wave patterns. *Amplitude* conveys power, such as the intensity of light or volume of sound.

Each memory wave has a unique waveform pattern, like a voiceprint. This diagram depicts a complex memory wave field composed of numerous individual memory waves.

Cognitive Engineering

4. Memory Particle

A memory particle is a conscious being's recording of a non-physical memory wave.

The memory particle carries a condensed representation of the information in the memory wave, including its frequency signature and amplitude.

The idea of memory particles may sound unfamiliar or arbitrary, but at one time, so did the idea of atoms. [4]

A memory particle has two states: inert and active.

4A. Inert Memory Particle

Memory waves condense into inert memory particles when they are recorded. Each inert memory particle carries a non-physical representation of one memory wave.

An inert memory particle is a memory wave recording which does not contain an awareness point. Inert memory is pre-unconscious, meaning it has the potential for becoming actively unconscious, but it is not presently unconscious. While a memory particle is in an inert state, it does not absorb any of the being's awareness resources, and cannot be recalled.

Almost all memory particles are inert. Only a tiny fraction of a being's total memory particles are ever active at one time.

---

[4] Atomic theory arose only recently in human history from John Dalton's work two centuries ago.

4B. Active Memory Particle

An active memory particle is an inert memory particle which has been activated by the application of an awareness point.

When an awareness point is applied to an inert memory particle, the particle changes state into an active memory particle.

The active memory particle emits a replica of the memory wave which it originally recorded.

When the inert memory particle is animated by an awareness point, the memory wave is reconstituted by the life energy of the applied awareness point and emitted. The applied awareness point is the force of the memory wave's projection.

5. Memory Record

A memory record is a collection of memory particles which are associated in time.

The memory record of an experience contains all perceptions – visual, auditory, spatial orientation, temperature, sensation, taste, smell, tactile, thought, emotion and so on. Each perception is encoded in and carried by a separate memory wave.

In the memory record for an experience, each memory wave has been recorded in a separate memory particle. One memory record contains the memory particle recordings of many memory waves.

In total, the memory record is a template or blueprint for re-generating the memory waves it originally recorded, like a movie film frame.

The memory record is not alive per se. Life energy is in the awareness points and the memory waves.

A memory record has two states: inert and active.

Cognitive Engineering

5A. Inert Memory Record

An inert memory record is a recording of experience which is comprised of inert memory particles. It is in a pre-unconscious state.

5B. Active Memory Record

An active memory record is a recording of experience containing active memory particles which emit memory waves. It is in an unconscious state.

Cognitive Engineering

● Summary

Memory attributes of the three major states of consciousness are summarized in Table 2 and explained below.

State	Points	Particles	Records	Waves
Conscious	Free Awareness Points	-	-	-
Unconscious	Applied Awareness Points ➡	Active Memory Particles ➡	Active Memory Records ➡	Memory Waves
Pre-Unconscious	-	Inert Memory Particles ➡	Inert Memory Records	-

– Table 2 –

Conscious: *Free awareness points* comprise the main body of conscious awareness.

Unconscious: *Applied awareness points* animate *active memory particles* in *active memory records*, which generate *memory waves*.

Pre-Unconscious: *Inert memory records* contain *inert memory particles* which are recordings of memory waves.

The dynamic interaction of these elements is discussed in the reference section's chapter on memory behavior.

Cognitive physics unlocks new worlds of awareness.

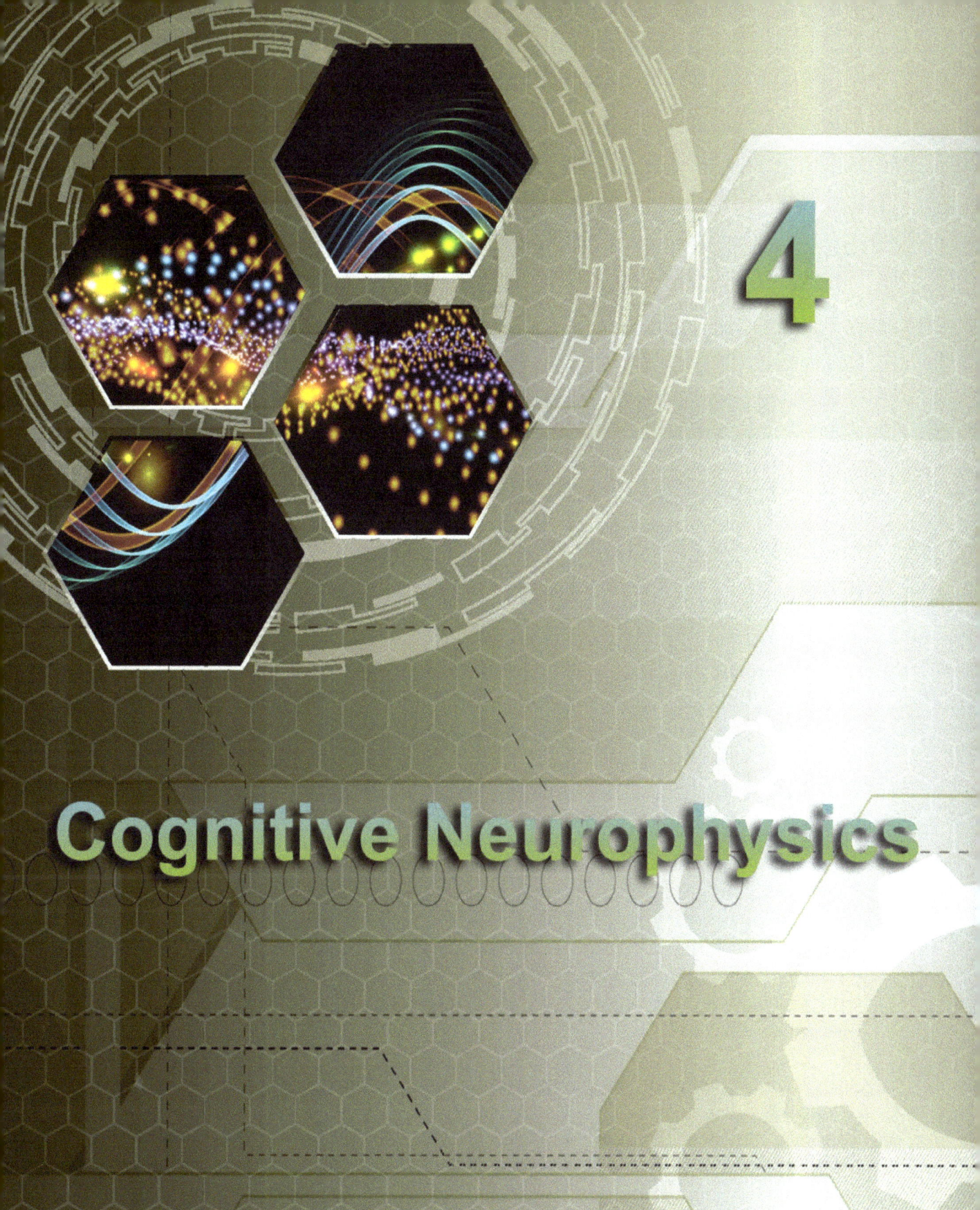

Cognitive Neurophysics

● New Sciences

This section introduces two new scientific disciplines which codify the relationship between the conscious being and the neurology it inhabits. These two disciplines form the platform for a cognitive neurogenetics science for enhancing human awareness.

Macro Memory (Unconsciousness)		
Discipline	Neurophysics	Neurodynamics
View	Micro	Macro
Memory Attribute Considered	Frequency	Amplitude
Memory Wave Function	Information	Power

Macro memory is composed of memory waves having two attributes: *frequency* and *amplitude*. Frequency is the universal carrier of information. Amplitude conveys power.

Two new sciences, *cognitive neurophysics* and *neurodynamics*, examine macro memory waves from the perspective of frequency and amplitude. Neurophysics studies the waves at a detailed micro level, while neurodynamics considers them at the whole-brain level.

Cognitive neurophysics studies the electrical relationship between awareness and neurology. It examines how electromagnetic waves act as the interface zone between the conscious being and its physiology, and it defines the properties of these waves in relation to different states or densities of consciousness.

Cognitive Engineering

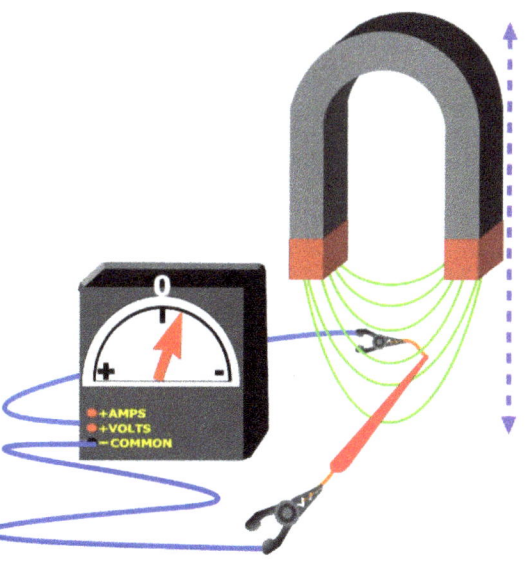

The upcoming discussion of neurophysics will require readers to understand two basic principles of electricity: *magnetic induction* and *electromagnetic radiation*.

1. Magnetic Induction
When electric current is traveling through a wire, we say that electrons are flowing through a conductor. There are several ways to make electrons move. One way to make electrons flow is with magnetism.

The horseshoe magnet in the diagram has a magnetic field which is represented by the green lines beneath it. Moving the magnet up and down, as shown by the purple dotted line, causes the meter to register an electric current.

The green magnetic field is positively charged (+). The electrons in the wire are negatively-charged (−). The positively-charged magnetic field attracts the negatively-charged electrons, causing them to flow through the wire. The resulting electric current registers on the meter. This principle is called *magnetic induction*.

2. Electromagnetic Radiation
In magnetic induction, magnetic fields make electrons flow. The converse of this principle is also true, which is: Flowing electrons generate magnetic fields.

As illustrated in the diagram, the flowing electric current **I** marked in red produces a magnetic field **B** shown in green. This principle is called *electromagnetic radiation*.

A common example of this principle in operation is power line radiation. The electrons flowing through high-voltage power lines generate strong electromagnetic fields. The same principle applies to household appliances and cell phones. However, since these devices have lower power, they generate weaker magnetic fields.

Cognitive Engineering

● Summary

Two fundamental laws of electricity are essential for understanding neurophysics. They are:

1. Magnetic Induction: Moving magnetic fields cause electrons to flow in conductors. This principle is known as *Faraday's Law*.

2. Electromagnetic Radiation: Electrons flowing in conductors produce magnetic fields. This principle is called *Ampere's Law*.

● Faraday's Law

Understanding the principle of magnetic induction is essential to grasping the concepts of neurophysics, so we will explore it in more detail.

Faraday's Law of Magnetic Induction is one of the fundamental pillars of electrical engineering. It explains how a moving magnetic field (like a magnet) causes electrons to flow in a conductor (like a wire). The positive charge of the magnetic field (+) attracts the negatively-charged electrons (-) causing them to flow in the conductor as electricity.

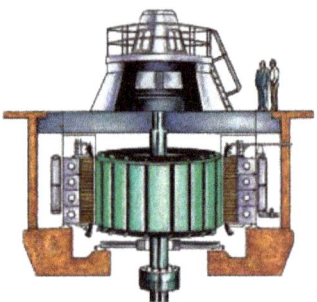

Example: Hydroelectric power generating turbines like the ones shown here at the Hoover Dam use magnetic induction to produce electricity. Water power from the dam spins giant magnets called *rotors* (shown in green) at high velocity. The rotors spin inside massive coils of wire called *armatures* (shown in brown) to produce a strong flow of electric current.

Although hydroelectric power stations are gigantic, a small turbine the size of a Volkswagen could produce electricity for powering a single home and its electrical appliances and computers.

Cognitive Engineering

Now imagine a miniature turbine the size of a basketball. Instead of generating electric power, though, this generator produces a life force which powers the physical processes of life in a human body, as well as the person's entire memory.

Think of a conscious being as the central magnet in the power generator, and the coils of wire as the brain's neural networks.

When a conscious being incarnates, it enters into an electrical circuit with the body's nervous system. The being's unconscious electromagnetic wave field impinges on the body's neurology, and it propels electrons through neural networks in the brain, just like rotors in power generators cause electricity to flow in armatures. These electrons continue their journey from the brain out through the central nervous system, eventually becoming nerve impulses which move muscles and animate the body.

Propagation Layers

Communication between an embodied conscious being and its neurology is governed by Axiom 2 (Condensation), which states that consciousness condenses through a series of layers to form the living material world. A corollary of this axiom, defined below, explains how embodied life is a continuum of several sandwiched layers which resonate together.

Corollary 3 – Propagation
Consciousness condenses through six layers to form the living universe.

Axiom 2 states the universe is a condensed form of consciousness. Corollary 3 amplifies this principle, showing how consciousness condenses in a cascade effect through 6 stages to form life.

Conscious beings deploy frequency signals to exchange information with their physical bodies. As shown in the diagram, *active memory particles* emit *memory waves* (gold) which are propagated through *space* (blue) and generate identical patterns in voltage fields (green) marked n*eurowaves*.

Neurowave voltage, in turn, propels *electrons* through *neurons*, powering the physical functions of life. Each of the layers illustrated in the diagram is explained in the pages which follow.

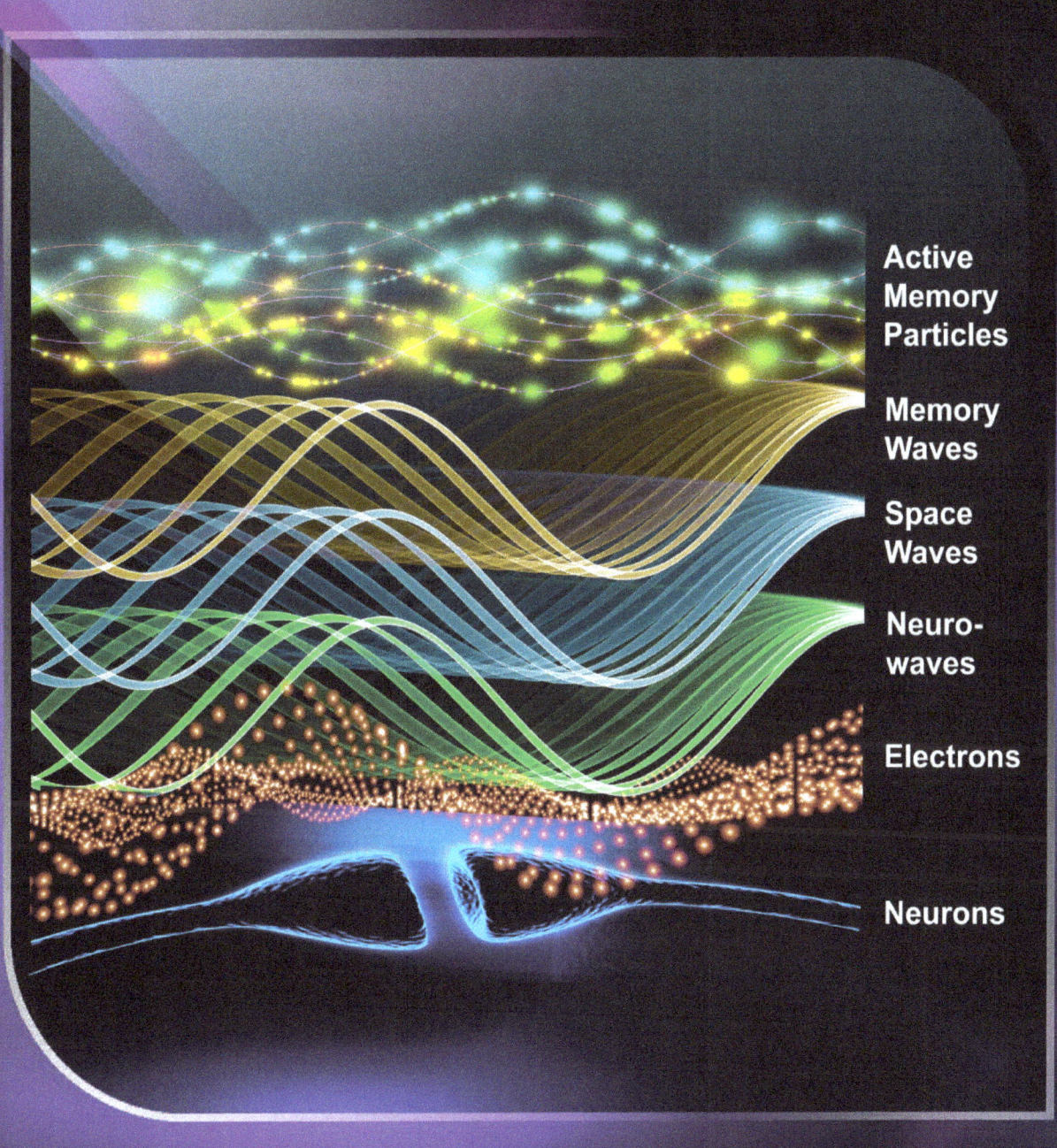

Cognitive Engineering

● Layer 1 – Active Memory Particles

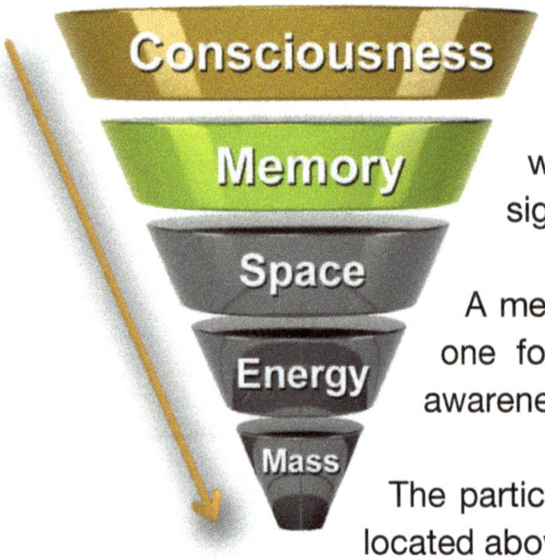

The luminous particles illustrated here are memories which have been activated by the application of awareness. They contain no mass, energy or physical wavelength; only a recording of the memory's frequency signature and amplitude.

A memory particle is like a transducer converting energy from one form to another. It converts the energy of an applied awareness point into the energy of a memory wave.

The particles behave as if they are floating on a plane of memory located above the dimension of space.

This field is the source of all the other layers shown in the diagram on the facing page. By generating unconscious memory waves, it forms the basis for memory functions and embodied life.

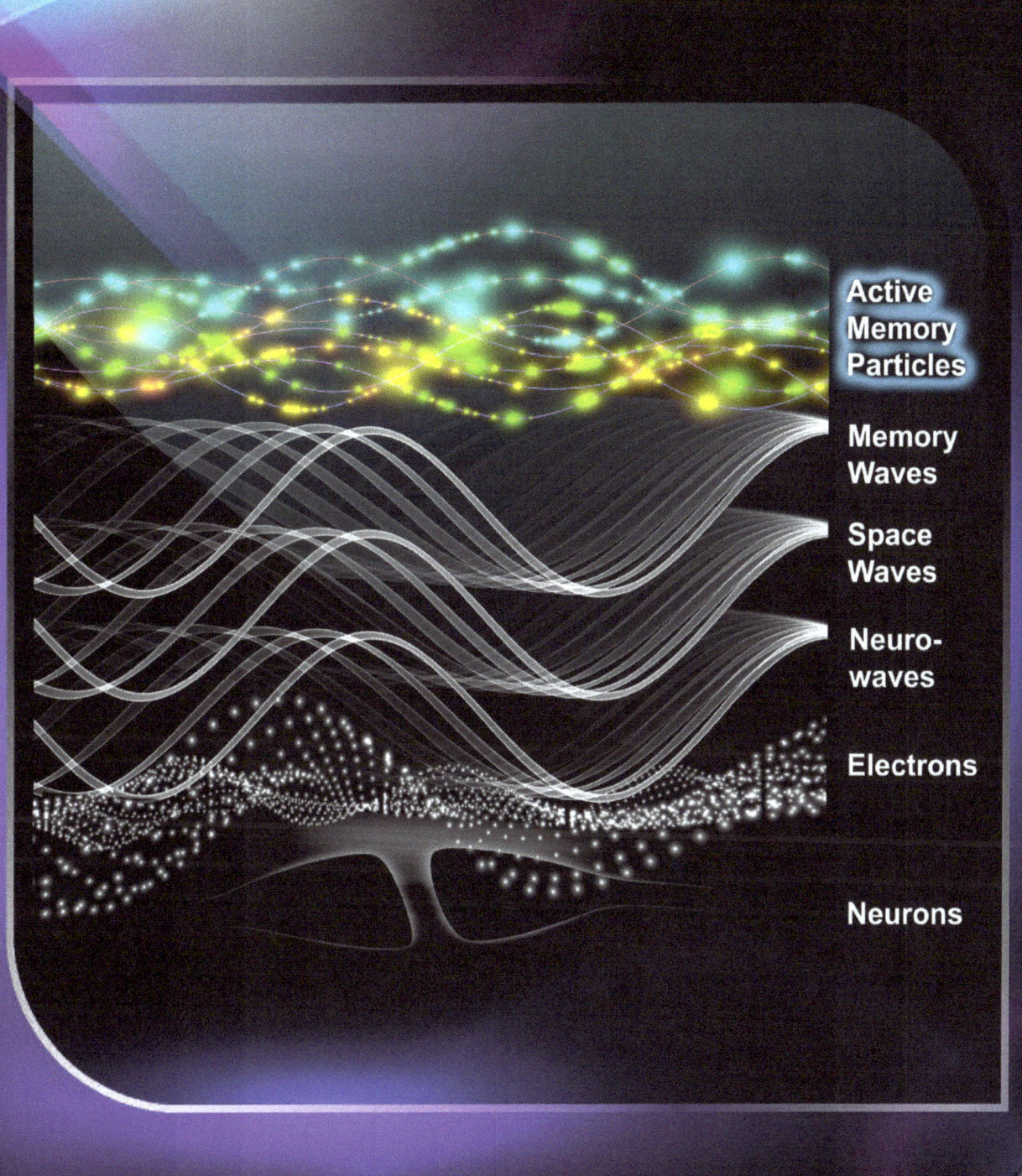

Cognitive Engineering

🟢 Layer 2 – Memory Waves

Definition

The energy of the universe is expressed in waves, including light waves, sound waves, and brainwaves. Every experience has a unique energy wave signature.

Awareness interfaces to the physical universe through waves. A conscious being perceives, communicates and remembers via memory waves. A being's memory waves are the storehouse of its knowledge.

Awareness in its unconscious state oscillates in complex wave patterns called memory waves. Memory waves store knowledge of internal states and external perceptions. Taken together, memory waves comprise the main body of unconsciousness.

The only reason there are past life memories is because conscious beings take this wave field with them wherever they go.

Characteristics

A memory wave holds the full array of perceptions and characteristics of the memory. Every memory wave has a unique frequency signature. Sine waves are shown in the diagram on the facing page for simplicity, but the actual waves are complex, like the voice print illustrated on this page.

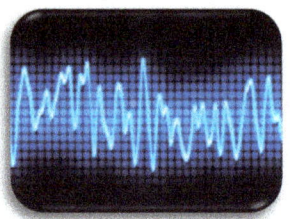

Memory waves contain no matter, energy, or space, but they do possess a non-physical frequency and amplitude. This "mental energy" is formed by the condensation of free awareness.

Recalling the song of a dove from memory generates the same pattern of brainwaves as hearing the bird song originally. If the information came in as a wave, and is recreated from memory as a wave, then it must be stored as a wave.

The memory wave draws its power from applied awareness points. The life energy of the applied awareness points powers and permeates the wave. The memory has no life of its own; all life energy comes from the conscious being.

Formation

Consciousness is formed by free awareness points which coagulate, touch each other, and resonate in a field. When awareness points leave the main body of consciousness to flow outward into memory waves, they are still awareness, but they are no longer contributing to the aggregate total consciousness. (Instead, they contribute to the aggregate total unconsciousness.)

Memory waves are formed by the condensation of free awareness. The memory wave is the first and lightest condensed form of awareness (unconsciousness). The memory wave contains awareness in an unconscious state (i.e., applied awareness). If the being is incarnated, the wave will condense further into space and energy.

Pure information in consciousness condenses into a memory wave form to insert it into the being's unconscious memory field and allow it to be preserved. The memory wave uses frequency and amplitude modulations to record the information.

The applied awareness point is a continuing source of energy like an electric power outlet. As the awareness point occupies a memory particle, it animates the particle's wave recording and regenerates the recording into a live memory wave. In animating the memory recording, the applied awareness point oscillates according to the frequency signature blueprint stored in the recording. The memory wave remains active as long as awareness is applied to it.

> Our intentions, attitudes and beliefs arise from decisions we make. Without memory, we would forget these decisions shortly after we made them. Our past does not determine our future – our memories do. We cannot change the past... but we can change our memories.

Cognitive Engineering

● Layer 3 – Space Waves

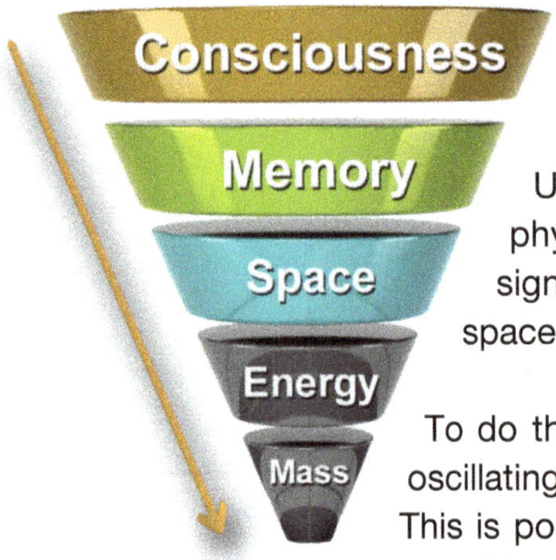

Overview

This layer marks the beginning of the physical universe.

Unconscious memory waves oscillate in unison with physical electromagnetic neurowaves. The memory wave signals are conveyed through an intermediate layer of space to the physical layer of electromagnetic waves.

To do this, empty space must be able to carry and propagate oscillating wave signals which have no electromagnetic energy. This is possible through zero-point energy; a concept universally accepted by scientists (originally developed by Max Planck and later modified by Einstein).

Zero-point energy is the energy of empty space. The fabric of space is a zero-point energy field which can be vibrated by frequencies which have no physical energy.

Zero point energy waves propagate in the vacuum of space and form overlapping interference patterns which carry information. Some leading physicists believe that zero point energy waves can actually carry memory.[5]

If this is true, then the zero point energy field is the bridge between consciousness and physical brainwave signals.

---

[5] *Science and the Akashic Field*, Ervin Laszlo, Inner Traditions, 2004

Exploring Space Waves

There is a boundary where a conscious being ends and space begins. It is impossible to detect this boundary in embodied states, but it is quite plain in very high angelic states.

A being in a high bodyless state is like a large soap bubble containing a hard vacuum surrounded by very thick air. The vacuum in this analogy corresponds to nothingness (awareness), and the air represents the space surrounding the being. Space is like a thick blanket of air around the being; much denser than awareness.

There is a border or membrane between the being and the surrounding space. Perhaps mind waves are fluctuations in the being's density which press on this membrane, like sound waves vibrate the eardrum to produce a corresponding neural signal. Vibrations in the awareness-space membrane, in turn, could generate a corresponding resonance in the zero-point energy field of space.

● Layer 4 – Neurowaves

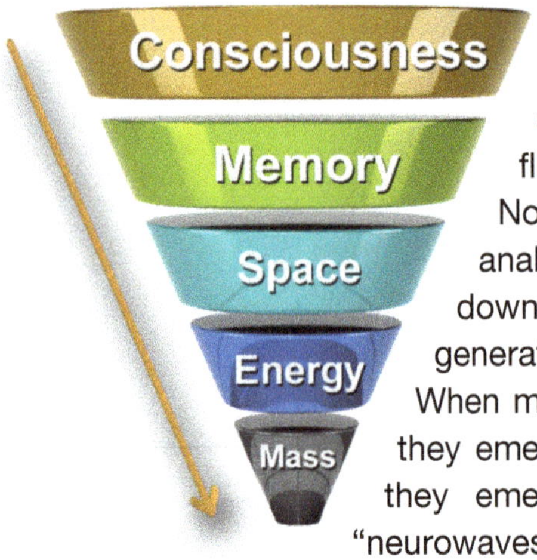

Overview

As wave oscillations in the zero-point energy field of space condense, they form a physical electromagnetic flux. Dr. Dean Radin, Chief Scientist at the Institute of Noetic Sciences, has suggested this condensation is analogous to the voltage reduction produced by a step-down transformer, in which a primary electrical field generates a secondary electromagnetic field of lesser strength. When memory waves first "step-down" into the physical realm, they emerge as space waves. When space waves step-down, they emerge as electromagnetic waves, which are marked "neurowaves" in the diagram and referred to throughout this book.

Neurowaves are tiny, cellular-level voltage waves. These fields are traditionally called brainwaves at the macro-level, but this text introduces the more precise term *neurowaves* to refer to microscopic electromagnetic waves at the neuron level.

This electromagnetic wave field impinges on neurology, and propels electrons through neural networks in the brain via the principle of magnetic induction (Faraday's Law).

The net effect of this cascading wave propagation is that when a conscious being incarnates, its unconscious wave component enters into an electrical circuit with the body's nervous system. The electromagnetic wave field in layer 4 of this propagation is the interface zone between awareness and physiology.

The amplitude of the neurowave is proportional to the amplitude of the memory wave at the top of the wave stack.

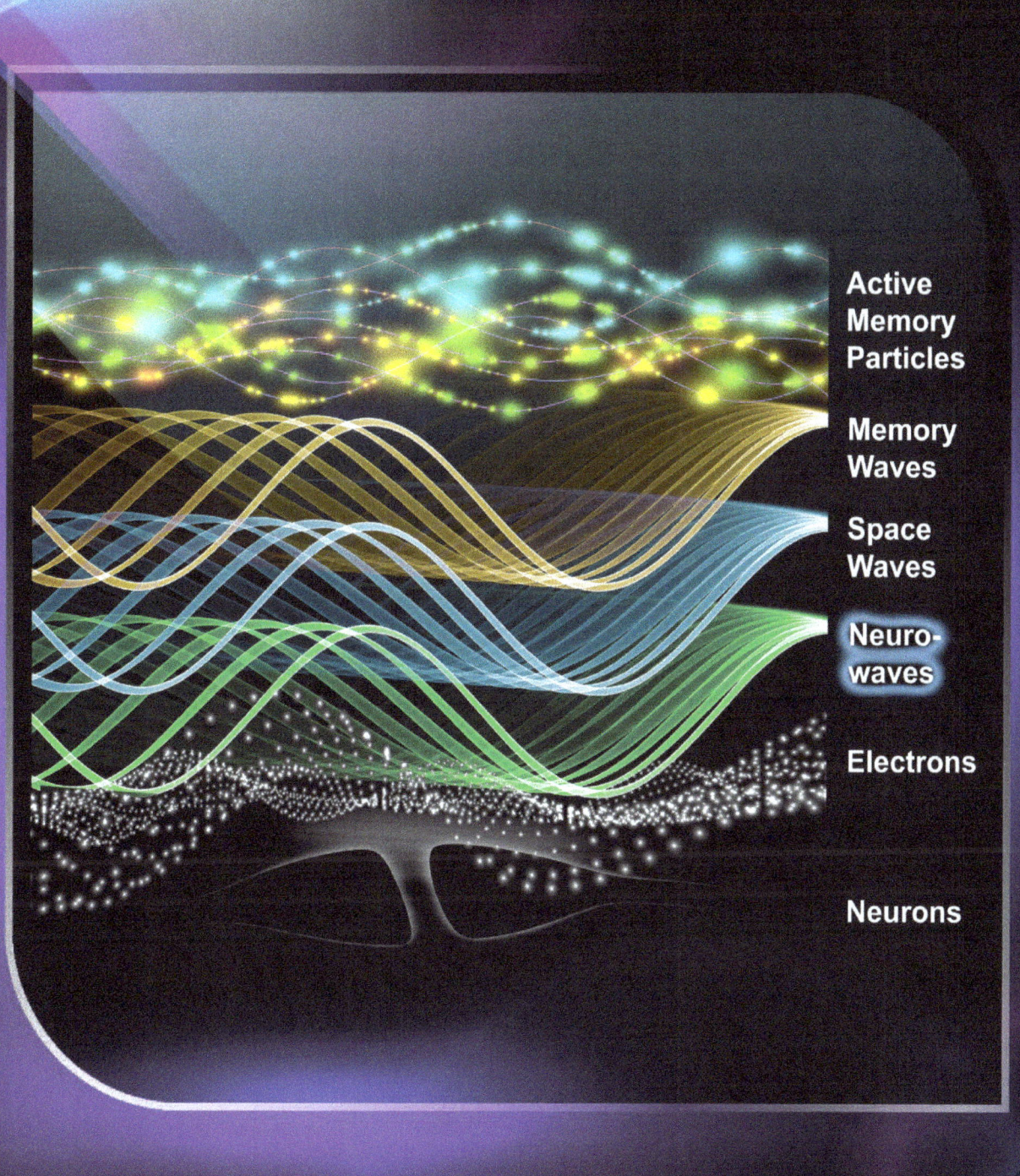

Cognitive Engineering

● Discussion

The basic building block of the brain is the neuron. From a functional perspective, a neuron is a switching mechanism for electrical impulses, much like a transistor.

Computers are made of transistors connected by wires, while brains are made of neurons connected by axons. In a computer, transistors act as logical switches which send electrical pulses along conducting wires. In the brain, neurons act as logical switches which send electrical pulses along interconnecting axons.

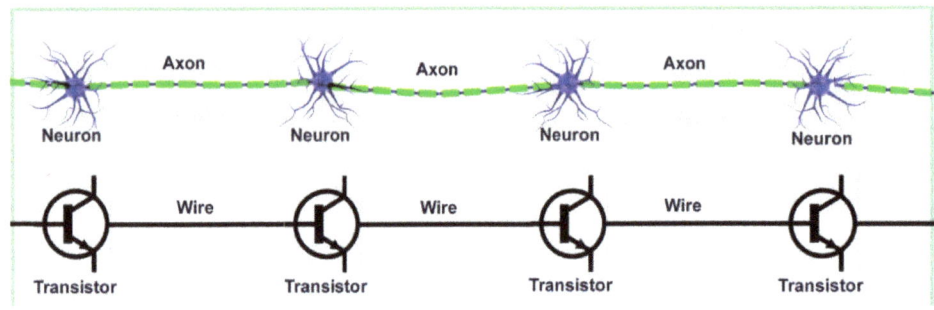

– Fig. 6 –

As shown in Fig. 6, the axon filaments (green) connecting neurons (blue) resemble wires connecting transistors in a series.

The performance of a microprocessor rests on its architecture and on the electrical characteristics of its basic building block, the transistor, or "gate." Change the speed of the gate and you change the whole performance of the computer.

Likewise, the brain's activity all rests ultimately on the electrochemical performance characteristics of the neuron.

● Cognitive Neurophysics

As stated earlier, research at Yale and Stanford has shown that flowing electrons in the brain's neural networks are accompanied by tiny electromagnetic waves measuring around 55 millivolts and 5 nanoamperes. (Relatively speaking, this is a lot of voltage to push a small amount of current. This large voltage is necessary to overcome the resistance of the brain's electro-chemical circuits, which is very high compared to ideal conductors like copper or gold.)

While flowing electrons in the brain generate electromagnetic waves, the converse is also true; i.e., electromagnetic waves can generate flowing electrons.

Cognitive Engineering

Electromagnetic waves have two different properties. They can *arise* from flowing electrons, or they can *cause* electrons to flow. The first case applies to *perception*, and the second to *behavior*.

Perception Example: When sound enters the ears, it is translated into neural impulses of *flowing electrons*, which *generate electromagnetic waves* that the conscious being can interpret.

Behavior Example: When an embodied being decides to speak, it emits *electromagnetic waves* which *propel electrons* in the brain and nervous system to move muscles in the body's larynx for articulating speech.

So neurowaves can arise from the body's senses and nervous system (perception), or from the being itself (behavior). While a conscious being is inhabiting a brain, the electromagnetic waves it emits propel electrons throughout the brain's neural networks.

● Frequency

Brainwave frequencies, conventionally expressed as a number between 1 to 40 Hertz, measure the average number of neuron conversations per second. When it takes longer for one neuron to talk to the next one, there are fewer neuron conversations in any given unit of time, and brainwave frequencies slow down.

While incarnated, the conscious being's electromagnetic field is electrically intertwined with its physical brain in the dance of life. The being's speed of thought, depth of perception, and other qualities are literally "gated" or limited by neuron speed.

Neurons and transistors alike transmit information as pulses of electromagnetic potential, or "voltage."

Before a neuron can send a pulse, it first must build up the energy for the pulse. Fig. 7 illustrates the time a neuron takes to accumulate this voltage, which is called *pulse rise time*.

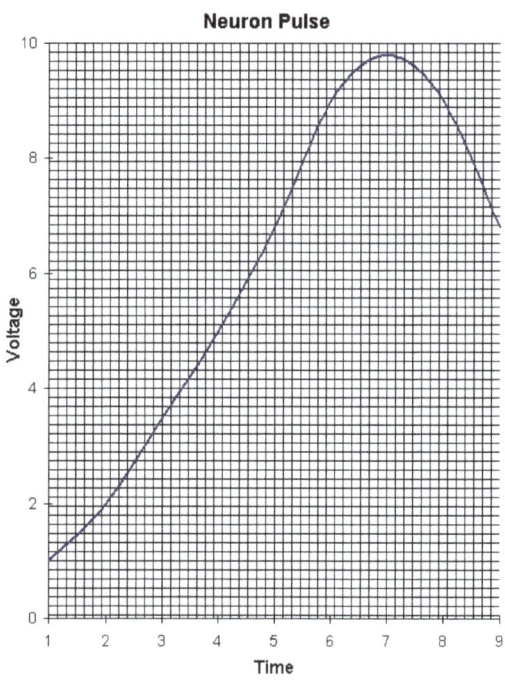

– Fig. 7 –

Once the energy in the neuron reaches the "threshold value" necessary to send a pulse (which is the top of the curve shown in Fig. 3), a spurt of energy is released from the neuron. This pulse is often called a neuron "spike," and its voltage is what brainwave measuring devices sense and convert into brainwave frequencies. For example, an average rate of 30 "spikes" per second would be reported as a brainwave frequency of 30 Hertz.

The "spike" of flowing electrons is transmitted from one neuron to the next one across the synaptic gap by way of neurotransmitter receptors.

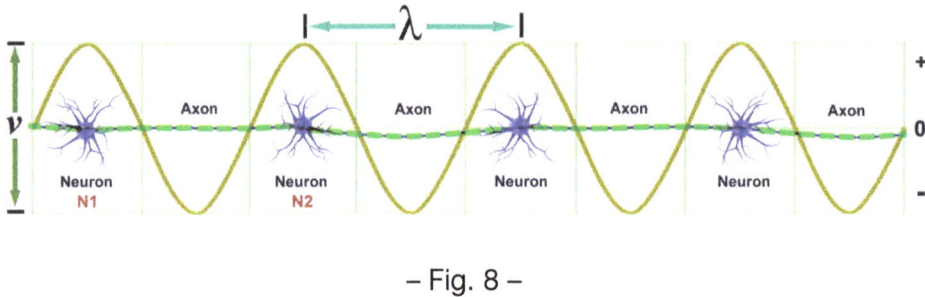

– Fig. 8 –

Figure 8 shows four neurons connected in a series by axons. Each neuron emits a pulse as shown in Fig. 8, which together form an electromagnetic wave or "neurowave." The neurowave shown in gold is plotted against voltage grid v.

As illustrated above, the neurowave's wavelength λ is equal to the time between peaks in the wave. This can be expressed mathematically as $\lambda = P+A$, where:
λ = Wavelength of neurowave;
P = Neuron pulse rise time; and
A = Axon transmission time.

In Fig. 8, the wave is energized when Neuron **N1** fires, then decays over the axon transmission until it is re-energized when the next Neuron **N2** fires.

Cognitive Engineering

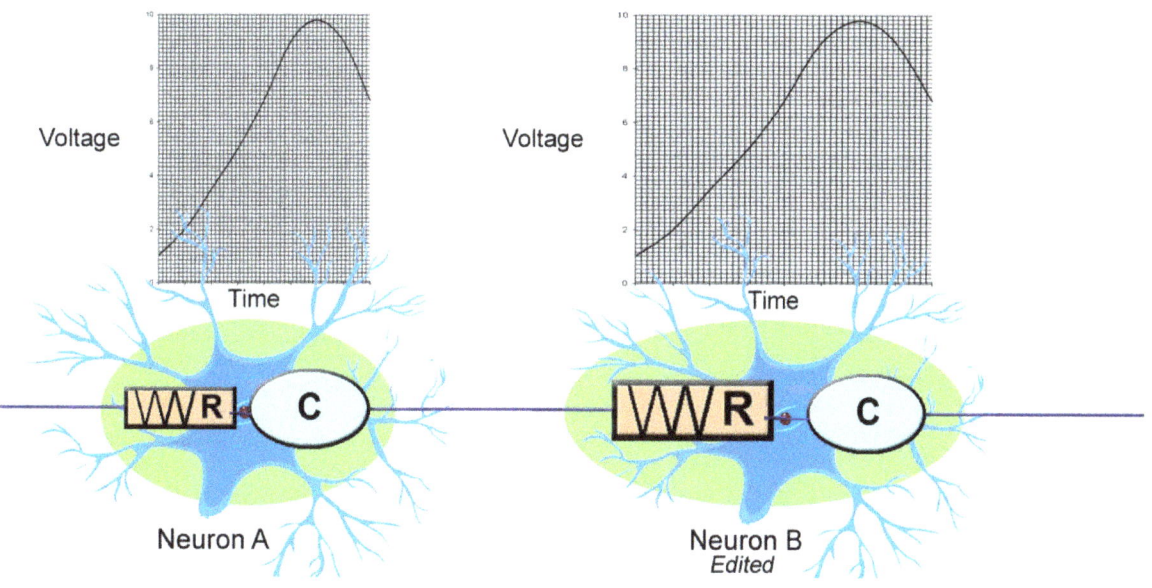

– Fig. 9 –

To understand how neurons generate electromagnetic waves, let's consider the neuron's electrical counterpart – the "RC circuit" (resistor / capacitor). In this kind of circuit, networks of resistors and capacitors are utilized to convey signal pulses made of electromagnetic waves.

A capacitor stores electrons which enter it like a reservoir holds water behind a dam. When the accumulated charge in a capacitor reaches its "threshold value," it discharges, and all the stored electrons in the capacitor flow over the dam, creating an electromagnetic pulse.

In an RC circuit, flowing electrons will enter a capacitor at a rate determined by the size of a resistor placed in front of the capacitor. A larger resistor will slow the electrons down; lengthening the amount of time it takes the capacitor to fill up.

The number of open input channels a neuron has to receive incoming electrons determines its resistance. The more open channels, the less resistance, and the faster it fills. The fewer open channels, the more resistance, and the slower it fills.

Neurons can be edited to raise their resistance by reducing the number of input channels, or "receptors" they have

Neurowaves are comprised of thousands of individual neuron pulses which are emitted by the neurons over which the wave travels. Slowing down even one of these pulses will change the frequency signature of the wave.

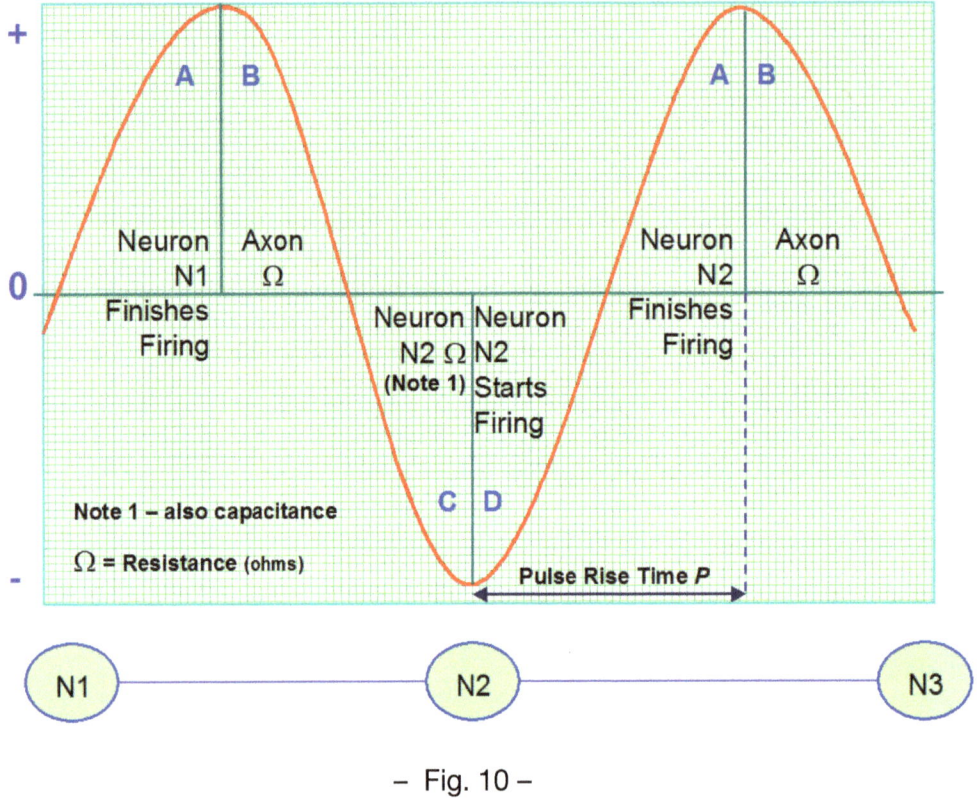

– Fig. 10 –

Let's take a closer look at the electrical characteristics of the neurowave by separating it into four quadrants: A, B, C, and D.

Quadrant A: Neuron **N1** releases its pulse signal at the peak of quadrant A in Fig. 5 above. The high voltage at the peak of the wave impels the signal across the axon.

Quadrant B: The signal's voltage diminishes in quadrant B above as it travels across the resistance of the axon.

Quadrant C: Negatively-charged electrons meet Neuron **N2**'s resistance, and gather in the capacitance reservoir of Neuron **N2**.

Quadrant D: Neuron **N2** begins to fire. The whole process repeats itself.

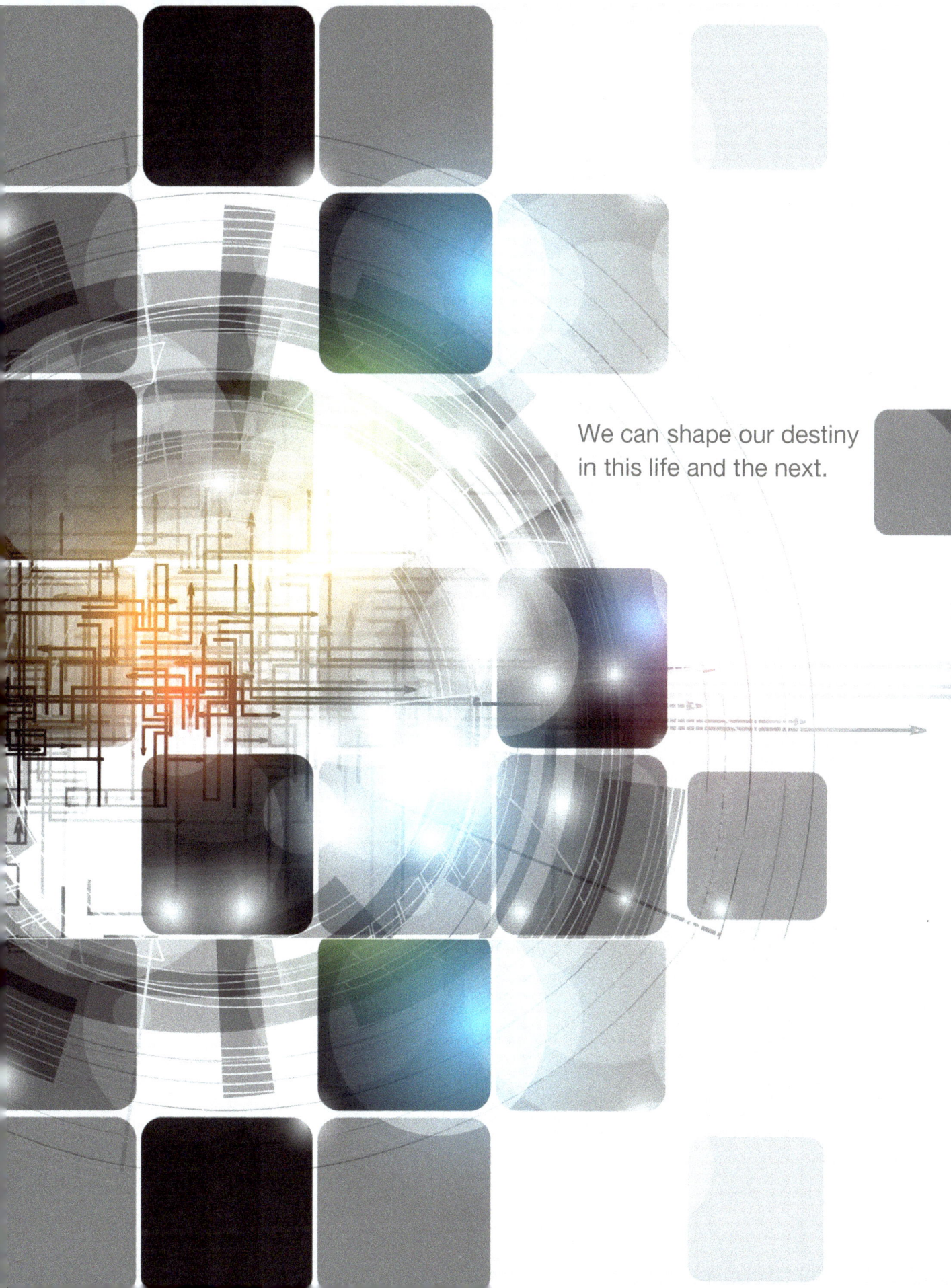

We can shape our destiny
in this life and the next.

Cognitive Engineering

● Layer 5 – Flowing Electrons

Electrons are matter which flows through conductors.

Electrons have a negative charge. The electromagnetic field has a positive charge. The positive charge of the field propels negatively-charged electrons, pulling them through axons and neurons in the brain.

In the brain, negatively-charged electrons are transported in atomic containers called *ions*. As illustrated here, ions (shown in gold) flow from one neuron to another across the resistance of the synaptic gap.

Moving electrons flowing through the body's brain and nervous system, power the functions of memory and physical life.

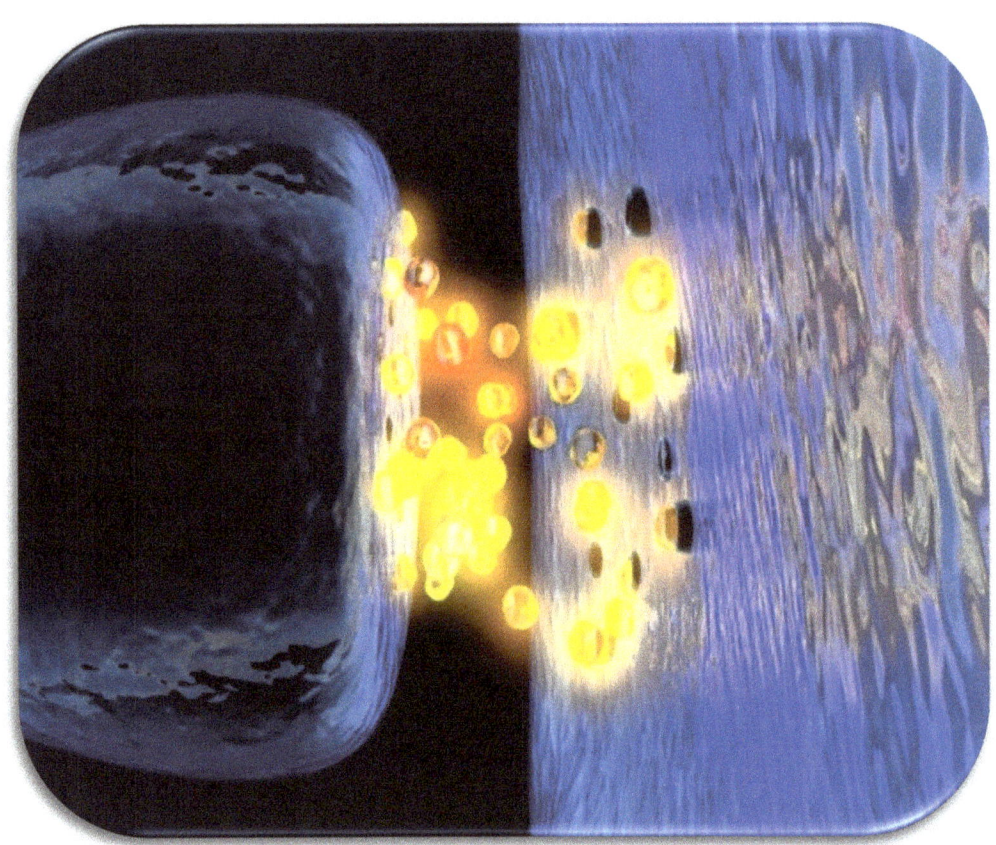

● Layer 6 – Neurons

The brain's conductors are neurons and axons. All conductors have resistance. The positive voltage of the neurowave's electromagnetic flux propels the negatively-charged, electron-carrying ions to overcome the resistance of the conductors, moving them through neural networks.

Beneath this level lies the realm of neurobiology, which is outside the scope of this book. The brain produces a wide variety of chemicals in response to mental activity. The flow of electrons through neurons is accompanied by a cascade of neurotransmitter molecules, primarily amino acids and peptides.

● Co-Residency

In the previous diagrams, particles, waves, electrons and neurons have been shown in parallel vertical planes for the sake of clarity. In actuality, they are co-resident, occupying the same volume of space, as illustrated below.

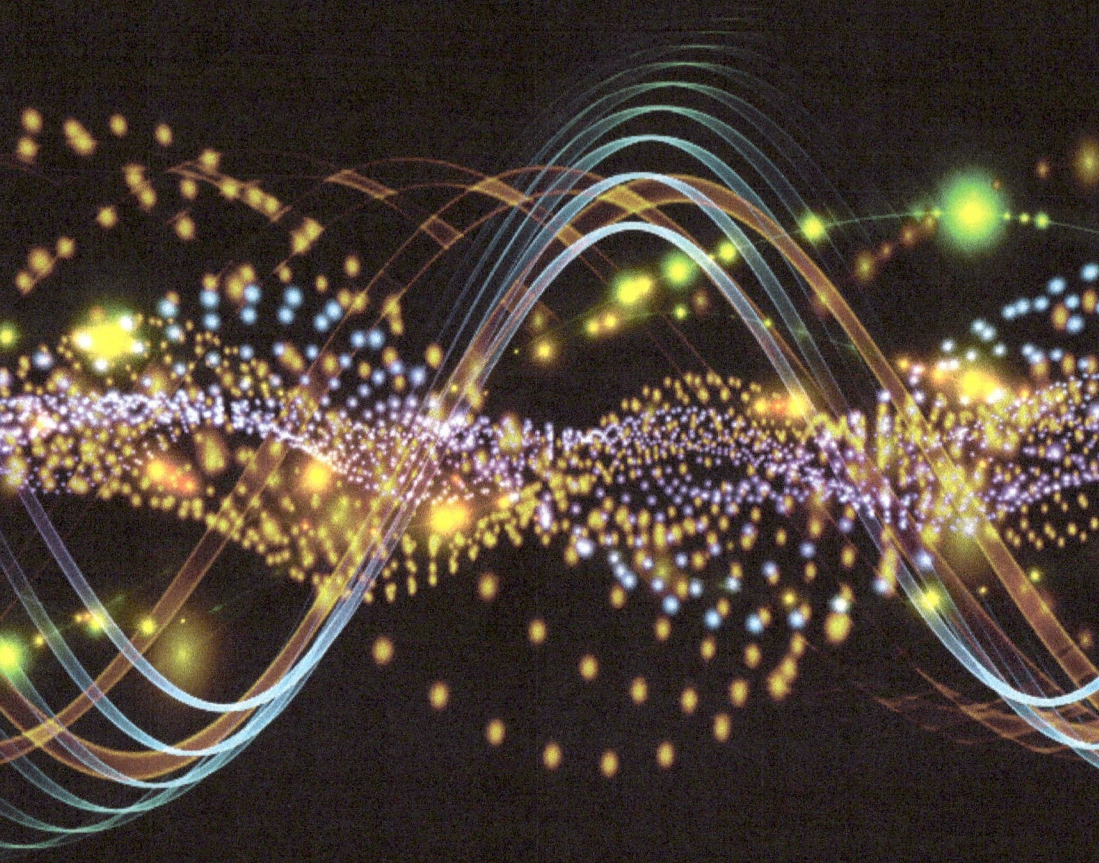

Cognitive Engineering

● Perception and Behavior

The principle of propagation has been illustrated in this chapter from a "top-down" perspective, starting with the conscious being and following the process of creation as it cascades through memory, space and electromagnetic fields to create effects in physical neurology. This process can be called *behavior*.

Communication between the conscious being and its body is a two-way street, encompassing both *behavior* and *perception*.

In *perception*, the six levels we have examined in this chapter are traversed in reverse. Nerve impulses from the body's sensory systems work their way up the ladder through neurowaves, space waves, and memory waves, until they finally arrive as perceptions in the conscious being.

This two-way avenue of behavior and perception is illustrated on the following two pages.

Behavior

Being
1. You decide to open a door.

Memory
2. You access your unconscious memory for the pattern for turning a doorknob. You emit a memory wave with that frequency signature.

Space
3. The memory wave enters the physical universe and is conveyed through space.

Neurowaves
4. The memory wave condenses into an electromagnetic wave.

Brain
5. The magnetic flux of the wave pushes against the electrons in the neural networks of your brain's motor cortex region, causing them to flow.

Body
6. Electrons flow from the brain through your nervous system to the nerves in your hand. Signals from the nerves instruct your hand muscles to turn the doorknob.

Induction

Perception

Read page from the bottom up.

6. You hear the bird song of a dove. → **Being**

5. The physical wave uncompresses into a memory wave. Your unconscious compares the memory wave with your audio memories for a pattern that fits the sound and finds one. → **Memory**

4. The electromagnetic wave is conveyed through space into the conscious being. → **Space**

3. Flowing electrons in the brain's auditory neural network generate an electromagnetic wave. → **Neurowaves**

2. The impulses arrive in the auditory area of the brain. → **Brain**

1. The sound wave of a bird song enters your body's ears. Your ears convert the sound wave into electrical impulses which travel along the auditory nerves. → **Body**

Radiation

Cognitive Engineering

● Perception

In perception, the brain works like a television broadcasting station and the mind works like a television set.

Television Studio	Person
Two television cameras	Eyes
Two boom microphones	Ears
Television broadcasting equipment	Brain
Television broadcast signals	Neurowaves
Television set	Mind
Television program	Unconscious activity
Viewer	Conscious being

Imagine there is a cat in a television studio. There are two television cameras in the studio and two boom microphones suspended above the stage. The cameras and the microphones feed streams of input into television broadcasting equipment. The equipment formats the input streams into television broadcasting signals, amplifies them, and sends the signals to a television broadcasting tower, where they are transmitted into space. (This example considers broadcast television which was in use in the early 21st Century.)

A television set receives the signal. The television's electronics are programmed to decode the signal. The signal tells the television set what color to make each one of the pixels on the television screen. It also refreshes the image on the screen 60 times per second. The result is a dynamic moving image of a cat in a studio.

Cognitive Engineering

🟢 Behavior

In behavior, the mind works like a movie projector and the brain works like a movie screen.

Movie Theater	Person
Movie projector's white light	Awareness
Movie projector machinery	Unconscious mind
Movie reel	Memory
Film frame	Memory record
Colored light film projection	Unconscious memory waves
Movie screen	Brain
Audience	Body

Think of the formless white light emitted by the projector as conscious, and the colored light projected through the film onto the screen as unconscious.

Memories can be thought of as recordings, like a hologram or a frame of movie film. The film frames on the reel are like inert memory records which carry every detail of the memory, like a hologram contains the full description of an object.

When white light is projected through a film frame, it becomes an active memory and emits a colored light projection which is like a memory wave. These waves superimpose themselves on neurology and the body. They can also enter consciousness. Memory recall adds a conscious observer to view the screen. Whenever the projectionist looks at the movie, the information on the screen enters their consciousness.

⬤ Laws of Perception and Behavior

The mechanics of a conscious being's incoming perceptions and outgoing behavior can be described by 8 laws which are illustrated on the facing page. As explained below, perception is governed by Ampere's Law of Radiation, and behavior is governed by Faraday's Law of Induction.

Perception
Read from the bottom up.

[Ampere's 4th Law for Awareness]
A changing memory field generates a changing awareness field.

[Ampere's 3rd Law for Memory]
A changing space field generates a changing memory field.

[Ampere's 2nd Law for Space]
A changing electromagnetic field generates a changing space field.

Ampere's Law
A changing electric field generates a changing electromagnetic field.

Behavior
Read from the top down.

[Faraday's 4th Law for Awareness]
A changing awareness field generates a changing memory field.

[Faraday's 3rd Law for Memory]
A changing memory field generates a changing space field.

Radin's Law
A changing space field generates a changing electromagnetic field.

Faraday's Law
A changing electromagnetic field generates a changing electric field.

Laws of Perception and Behavior

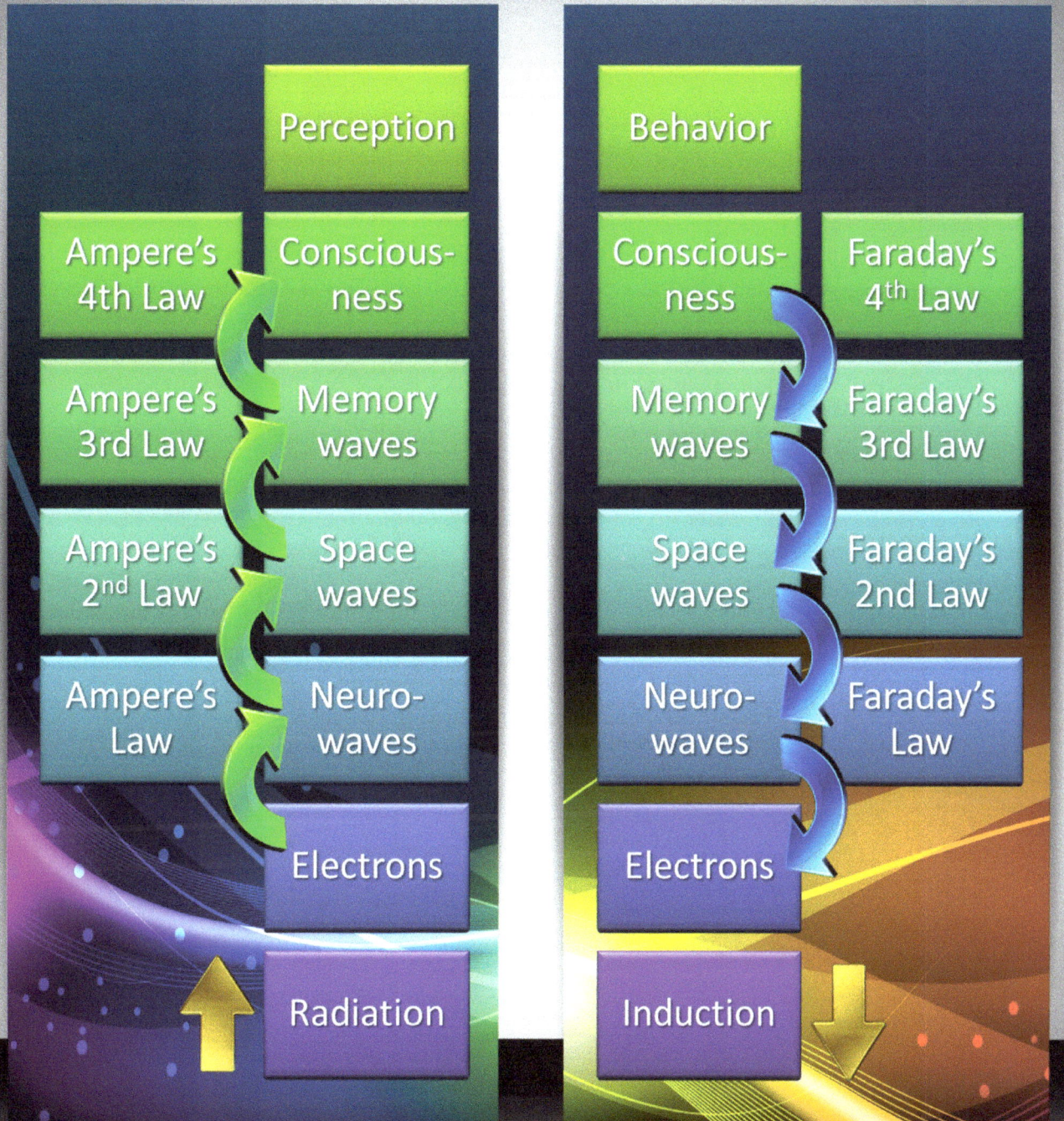

● Information Density

In this chapter on propagation, we have considered the transmission of information across 6 layers of mediums.

In Layer 5–Electrons, and Layer 6–Neurons, flowing electrons carry information through axons and neural networks in charged atomic containers called ions; matter moving through matter. This dense storage medium holds relatively little information per cubic centimeter and has slow electrochemical transfer rates.

In Layer 4–Neurowaves, we know that electromagnetic waves like television waves can carry concentrated information in fields which transit space at light speed.

From what is known about the zero point energy fields in Layer 3–Space waves, it seems that the information in this layer is even richer than in electromagnetic waves.

Memory waves are the least dense medium in the spectrum, and they would appear to carry the most information (particularly if they are holographic, as discussed later in this section). The final layer, consciousness, has zero density, and its storage capacity is almost infinite.

From these observations, one can deduce that as the medium becomes less dense, it can carry more information. If this is true, then information density is inversely proportional to the density of its storage medium. The denser the storage medium, the less information it can carry. The more uncompressed the storage medium, the greater information it can hold.

Cognitive Engineering

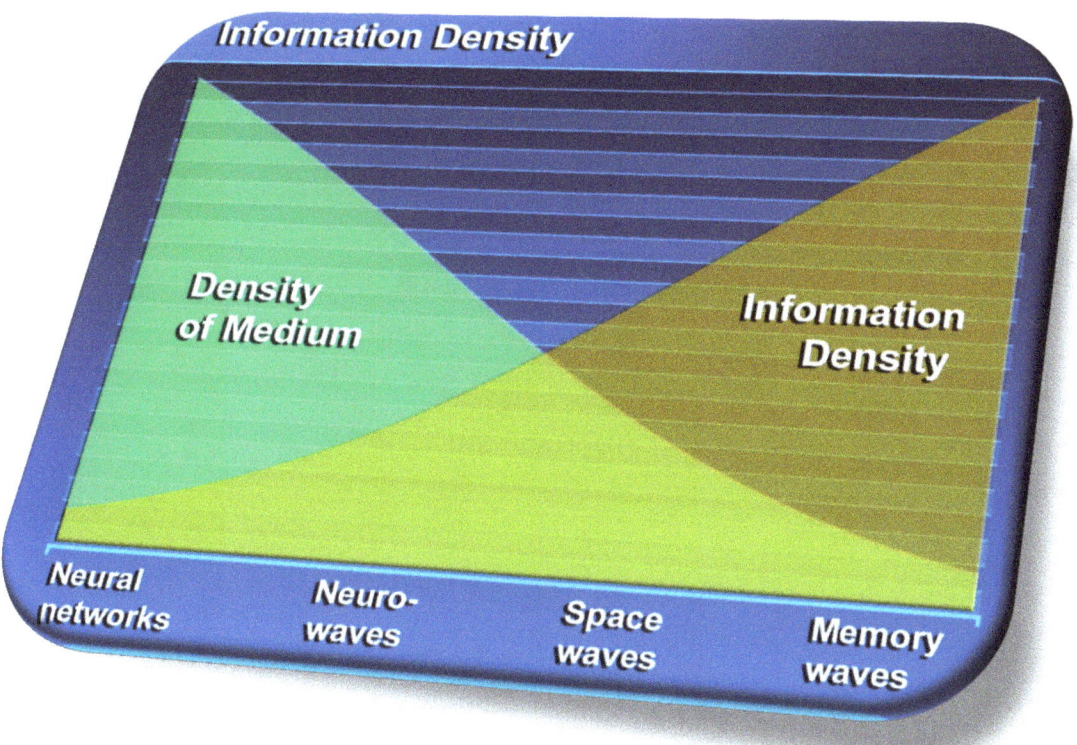

This relationship can be expressed mathematically as follows:

Formula 3
Information Density
Where:

$$\rho_i = \frac{1}{\rho_{sm}^3}$$

ρ = density
i = information
sm = storage medium

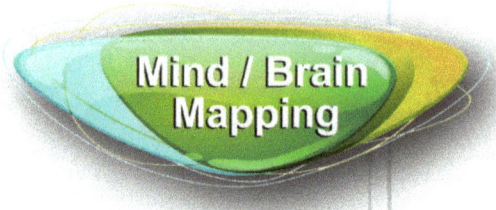

Mind / Brain Mapping

● **Introduction**

The preceding chapter has utilized two-dimensional illustrations and analogies in an attempt to achieve clarity in presenting a complex subject. The current chapter builds on this foundation and expands the discussion into three dimensions.

For the sake of simplicity, educators often represent electromagnetic waves as two dimensional lines (such as sine waves). This text has employed this convention in discussing wave propagation, as recapped in Fig. 11.

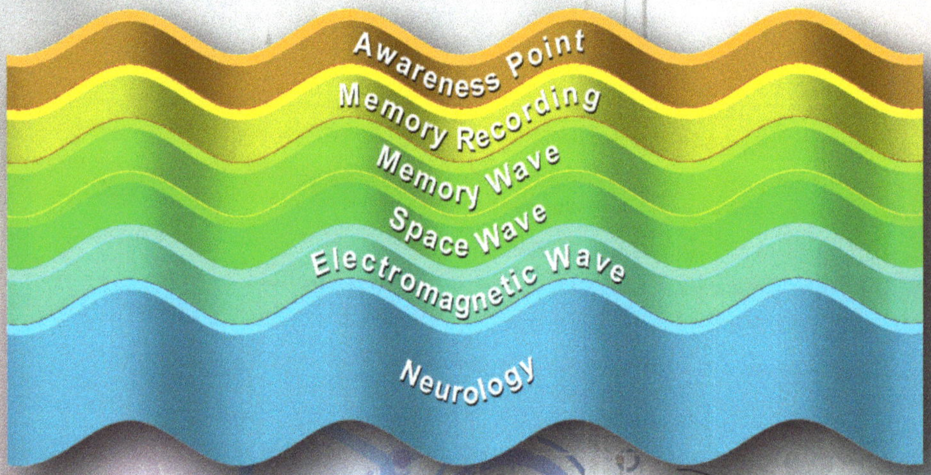

– Fig. 11 –

Reality is slightly more complicated. Electromagnetic waves actually flow in three dimensions, as shown here.

Viewed from three dimensions, wave propagation emanating from an individual active memory particle actually looks like the diagram in Fig. 12. In embodied beings, the living universe forms around applied awareness points in successive layers of time, space, energy, and matter.

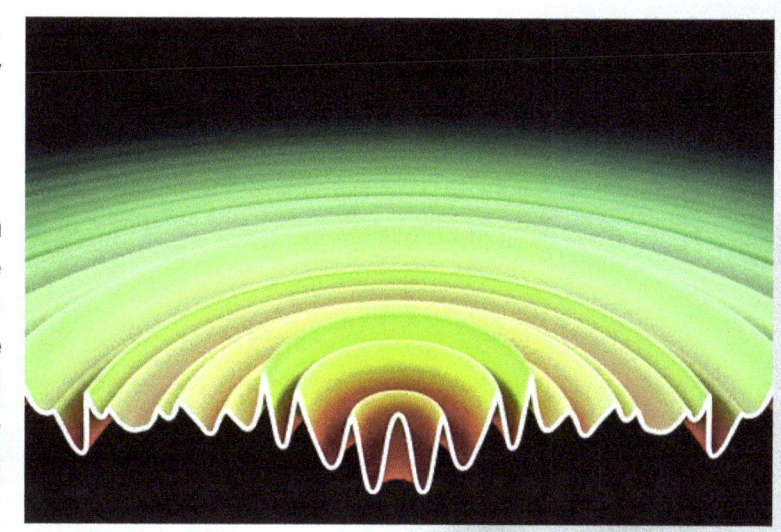

– Fig. 12 –

Free awareness points are not surrounded by time or form. These uninsulated points touch each other, coagulate into a unified body, and, if there are enough of them, resonate with self-awareness.

Per Axiom 1, applied awareness points have a form and are located in time. Form and time are like clothing which insulates an applied awareness point from the main body of free awareness points.

If the being is incarnated, the applied awareness point is surrounded not only by time, but also by space, energy and matter.

Each applied awareness point is the source of a neurowave projection. The radial propagation model in Fig. 12 shows how each awareness point can project into the physical universe, and also withdraw from it.

The model also works very well with mind/brain mapping because it explains the distribution of brainwaves across the neural topography.

● Logical / Physical Mapping [6]

The conscious being's unconscious memory field has a topography which organizes different types of memories into specific areas, like a database schema.

When this logical mapping is superimposed on physical neurology, it provides communications channels between the being's memory areas and their corresponding regions in the brain.

For example, the active logical memory record shown in Fig. 13 contains 5 memory particles A through E. Each particle holds the recording of a different type of memory wave. When activated, the 5 memory waves map to their corresponding areas in physical neurology.

---

[6] In computer engineering parlance, *logical* means *software* and *physical* means *hardware*. "Logical" as used here means "mental," as opposed to "rational."

Cognitive Engineering

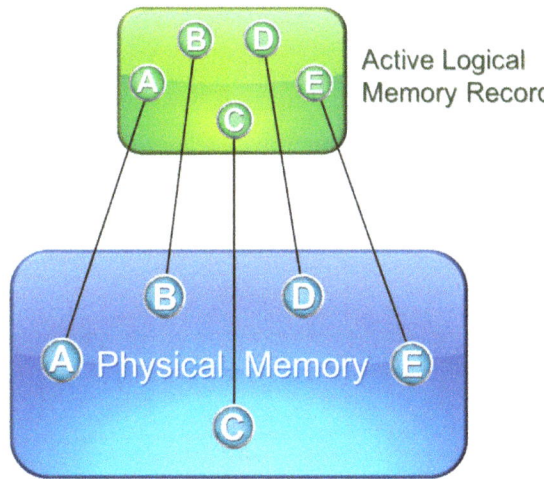

Area	Memory
A	Beliefs
B	Spatial
C	Emotion
D	Sensation
E	Visual

- Fig. 13 -

● Mental Energy Centers

The body stores the physical counterparts of a conscious being's memory in different specialized regions of the brain. Specifically, the physical memories of different perceptions are stored in the respective brain regions from which they arose. For example, visual memories are stored in the occipital lobe, and auditory memories are stored in the temporal lobe.

Similarly, the conscious being stores different non-physical memories in different regions of its unconscious mind. (As above, so below.)

The chart which follows shows the logical/physical mapping for 16 cognitive areas called *mental energy centers*. These centers contain different categories of mental activity which have unique frequencies.

Mental activity is on the left. Physical perceptions are on the right. Perception goes bottom up. Behavior goes top down.

Cognitive Engineering

Ⓐ Layer A – Conscious Awareness
A conscious being can be thought of as having a topography which experiences different types of cognitive activity in different areas. The top layer in the diagram represents a being's conscious experience, which has no frequency or amplitude.

Ⓑ Layer B – Unconscious Awareness (Memory Waves)
This layer contains memory waves with non-physical frequency and amplitude. Each of the 16 mental energy centers has a unique band in the memory frequency spectrum, like an area code covers a range of phone numbers (as discussed earlier).

Layer A is superimposed on layer B via the Propagation Corollary. As it condenses, cognitive activity in layer A manifests itself as memory waves in layer B. [7]

- -

↓ E m b o d i e d

Ⓒ Layer C – Space Waves *(not shown)*
Memory waves in layer B condense into space waves in layer C. Space waves have the same frequency organization structure as the memory waves they originate from.

Ⓓ Layer D – Voltage Waves *(not shown)*
Space waves in layer C condense into neurowaves in layer D, maintaining their original frequency organization pattern.

Ⓔ Layer E – Neurology
The neurowave voltage waves in layer D superimpose themselves on different areas of physical neurology based on frequency.

---

[7] Memory waves are actually complex frequency signatures, not one-dimensional wavelengths, but the chart shows the basic principle.

Mental Energy Centers

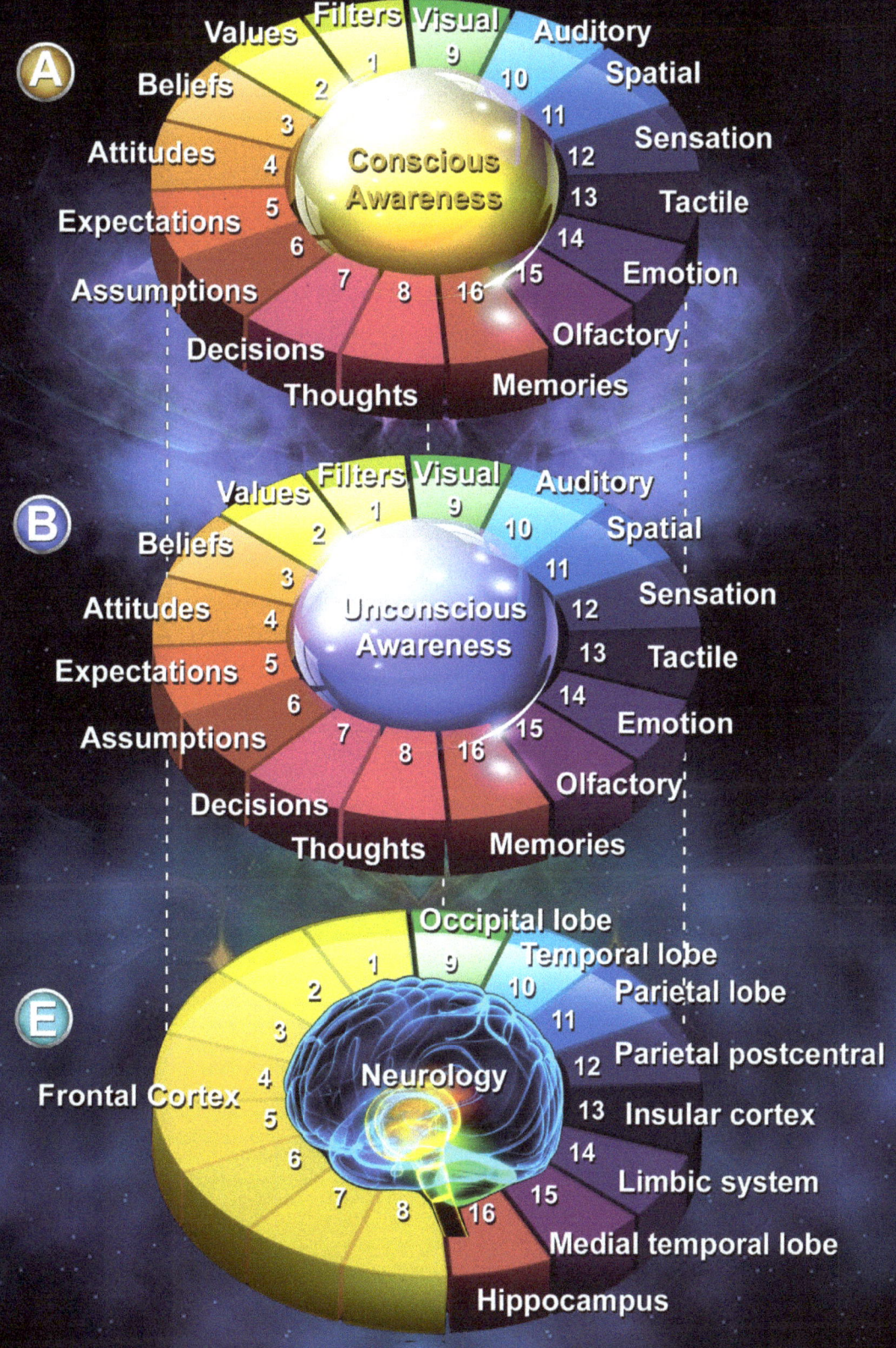

Memories are thought to cluster around 16 primary mental energy centers. These mental energy centers contain the 16 primary types of experiences. One of the mental energy centers is likely to be the predominant or largest factor in each significant memory.

Left Brain
1. Filters
2. Values
3. Beliefs
4. Attitudes
5. Expectations
6. Assumptions
7. Decisions
8. Thoughts

These 8 cognitive centers all map to the frontal cortex in the brain. They contain mental activity which has amplitude (energy) as well as frequency. Plain data has frequency but almost no amplitude, Therefore, factual knowledge is not a mental energy center and does not appear on the chart.

"Filters" are preconceptions which color our perception of the world. The other categories are self-explanatory.

Right Brain
9. Visual
10. Auditory
11. Spatial
12. Sensation
13. Tactile
14. Emotion
15. Olfactory
16. Memories

Different frequency signatures of neurowaves resonate and activate corresponding neural networks in physical regions of the brain. Areas in the right brain matching the 8 mental energy centers shown are indicated on the chart. The last area, the production of memories in the hippocampus, is treated in more depth in the following pages.

● Summary

The conscious being's unconscious memory field organizes different types of memories into specific areas. When the being's field impinges on neurology, the resulting mind/brain mapping topography corresponds to a "database schema" in computer parlance.

As this logical mapping is superimposed on physical neurology, it provides communications channels between the being's memory areas and their corresponding regions in the brain.

A conscious being's information "database" is organized into a hierarchy based on recency of access. Current information in present time is conscious, like the data in a computer's cache memory. Recently-accessed information is unconscious, like data stored on a disk drive. Inactive information is inert, like archive storage in a vault.

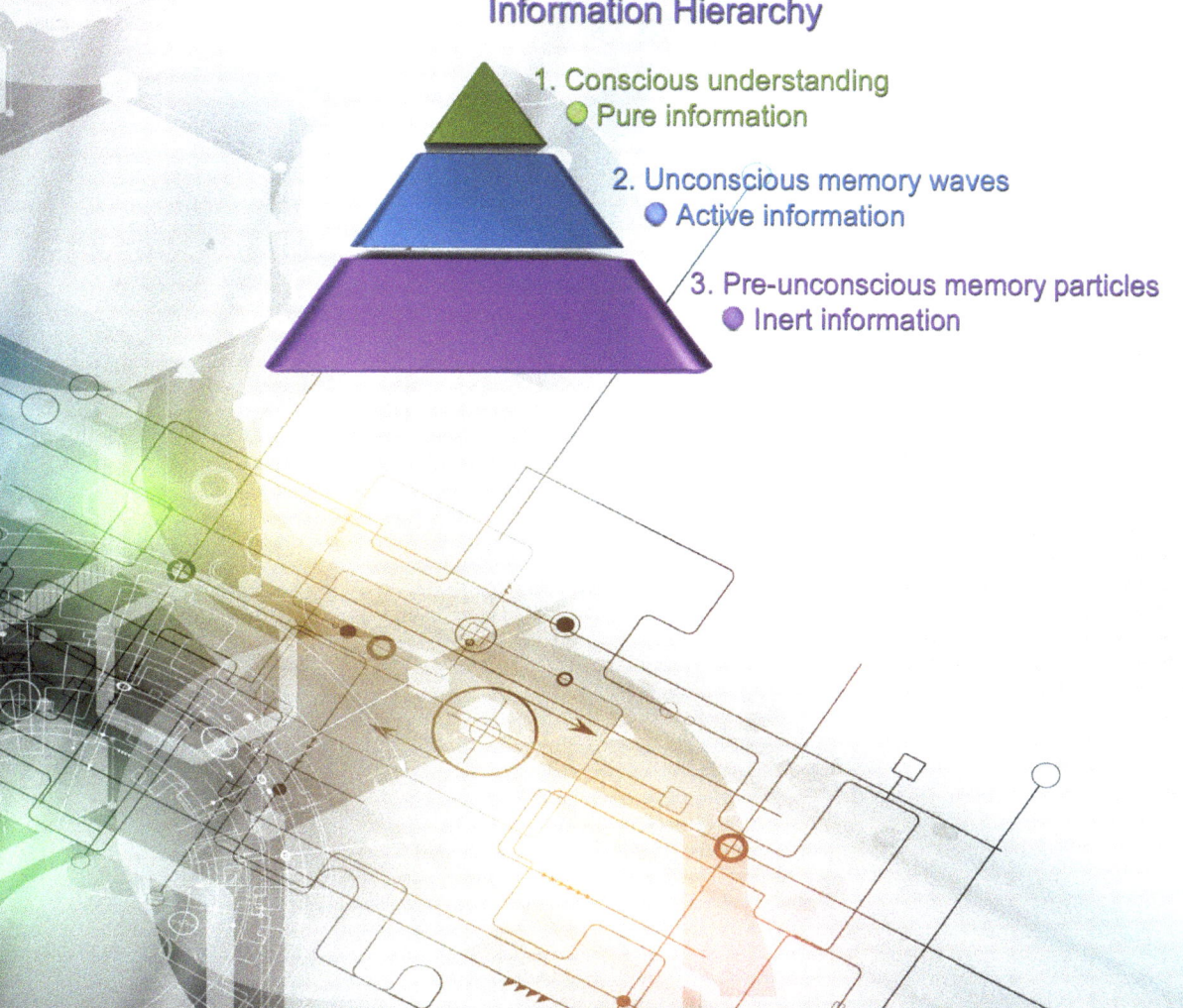

Information Hierarchy

1. **Conscious understanding**
   ○ Pure information

2. **Unconscious memory waves**
   ● Active information

3. **Pre-unconscious memory particles**
   ● Inert information

Engineering, science and mathematics can be harnessed to generate cognitive capital.

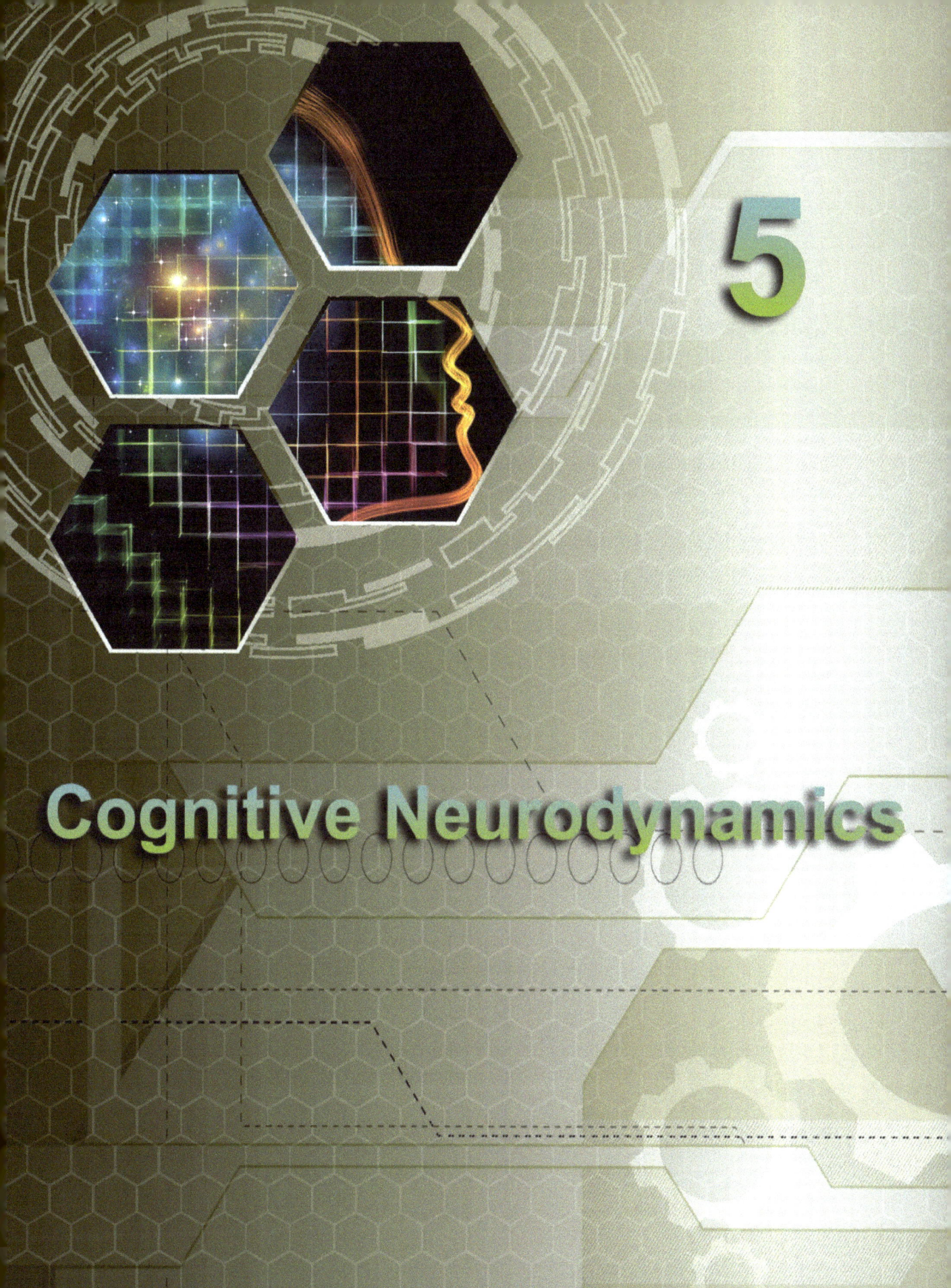

Cognitive Neurodynamics

5

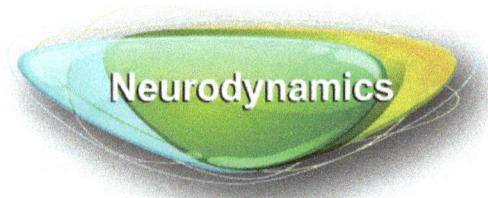

Neurodynamics

🟢 Introduction

The preceding chapters have explained how the conscious being generates *behavior* through electromagnetic *induction*, and how it receives *perception* through electromagnetic *radiation*. These activities are recorded in two kinds of memory: *behavior templates* and *perception records*.

Behavior templates, also called implicit or procedural memory, contain memories of central nervous system motor behavior patterns (e.g., how to ride a bicycle, how to speak, how to open a door).

Perception records, also called explicit or declarative memory, contain recordings of external perceptions (e.g., events and sensations) as well as recordings of thoughts (e.g., the meanings of words and objects, and identities of people).

A person needs to engage both types of memories in order to function in life. Memory is all-pervasive. People rarely perceive present time for what it is, without the overlays of memory.

Memories actively function without our conscious intervention to enable us to understand language, navigate the world around us, and recognize objects and people.

Most of the human mind is unconscious, along with most of our brainwave activity. Brainwaves primarily represent memory activity.

All active memories are unconscious. When a memory is remembered, it is no longer a memory; it is an element of consciousness.

Neurophysics and neurodynamics both examine how the incarnated conscious being communicates with its brain via memory waves. Neurophysics focuses on the *frequency* of the waves, and neurodynamics considers their *amplitude*.

Frequency, measured in Hertz, is the universal carrier of *information*. Amplitude, measured in Volts, conveys *power*.

Cognitive Engineering

● Perception Records

These two graphs depict the same perception memory at different amplitudes.[8] The frequency pattern of the wave is the same, but the larger wave carries more amplitude or power.

Walking on sand

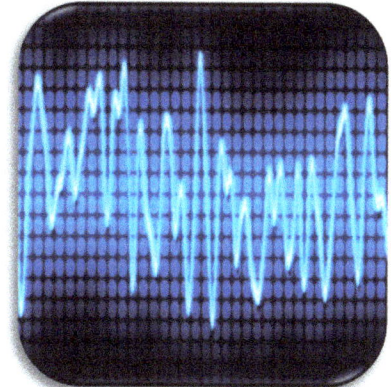

Walking on hot sand

Body sensations arrive in the brain at different amplitudes. Thus, their memory recordings have different amplitudes. For example, the memory of walking on hot sand has a higher amplitude than the memory of walking on normal sand. Hence, activating the memory of walking on hot sand requires more power than activating the memory of walking on normal sand.

If we compared these two memories to light bulbs, the normal sand would be like a 25 watt bulb, and the hot sand would be like a 100 watt bulb. Activating the vivid memory draws more power from the conscious being than activating the calm memory. (Conversely, deactivating a vivid memory, or recalling it into consciousness, releases more power than deactivating or recalling a calm one.)

---

[8] These are demonstrative examples, not actual memory recordings.

Cognitive Engineering

● Behavior Templates

Signals from the conscious being go out to its body at different amplitudes. For example, the memory of how to run is the same motion pattern for slow jogging as it is for fast running. However, racing requires more power.

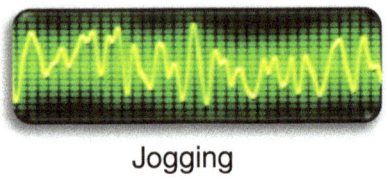
Jogging

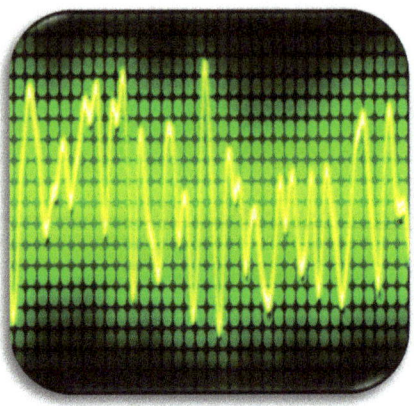
100 yard dash

When someone is running a 100 yard dash, their brain draws more awareness from their life force, just like an electric car draws more current from its battery when accelerated to its maximum speed.

From these observations, we can deduce the following:

Corollary 4 – Memory Power
The amount of awareness required to animate a memory is proportional to the memory's amplitude.

High voltage memories draw more power to animate than low voltage memories. This is true whether the memory is an internal memory or a behavior template.

Cognitive Engineering

● The Brain Appliance

The brain is like an electrical appliance that is powered by the conscious being's unconscious memory waves. It draws its power from the conscious being like a household appliance draws current from a power outlet.

When the conscious being applies greater numbers of awareness points to animating unconscious memories, the Law of Conservation of Awareness tells us there will be fewer free conscious awareness points left over in the conscious being.

Memories are like light bulbs. The bigger they are, the more power they consume. For example, if there are 100 watts of power available, and we use a 60 watt light bulb, there are 40 watts remaining in reserve. If we use a 90 watt bulb, there are 10 watts left over. Also, the more memories are lit up, the more total power they draw.

● Voltage

The amount of power the brain draws depends on the amplitude (voltage) of the active memory waves. High voltage waves draw more power than low voltage waves, just like a 90 watt bulb draws more "juice" than a 60 watt bulb.

Conversely, when the brain draws less life force, such as during relaxation and meditation, then there is a greater amount of free conscious awareness available.

To summarize:
- High voltage memories require more power to animate.
- Low voltage memories require less power to animate.

Cognitive Engineering

● Experiments

Neuroscience experiments uniformly demonstrate that subjects in a resting state generate a large amount of brainwave activity. If the subjects are not consciously engaged in any mental or physical activity, where are all these brainwaves coming from? To find the answer to this question, consider the following illustrations.

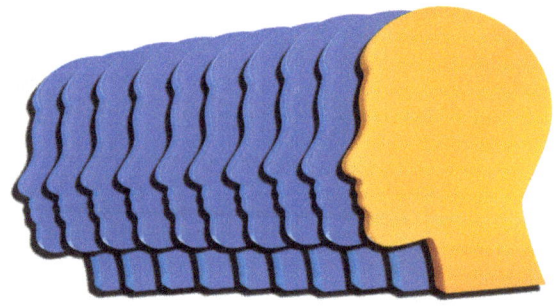

Suppose you are ten conscious beings instead of one. Nine of the conscious beings who comprise you are asleep, and one is awake. The ones that are sleeping are not inert; they are all dreaming, and dreaming vividly. Their dreams produce brainwaves which reflect unconscious activity.

Experiments indicate that when one of these nine sleeping people wakes up, the overall brainwave activity goes down. So now you have two people awake, and eight people asleep, and fewer brainwaves.

This moves you from being 10% awake (position 10 on the periodic table of awareness) to being 20% awake (position 20 on the table). Position 20 has twice as much free conscious awareness as position 10, and hence, reduced unconscious brainwave activity (per the Law of Conservation). Conscious awareness has no wavelength. Unconscious awareness does.

The Propagation Corollary tells us:

1. The more applied awareness points there are, the more memory waves there are (in both number and amplitude).

2. More memory waves give rise to more space waves.

3. The more space waves there are, the more neurowaves there are.

4. Greater numbers of neurowaves cause more electric current to move through neurology.

To summarize, the more applied awareness points there are, the more electric current is moving through the brain.

Applied awareness points are awareness in an unconscious form. So if the above laws are true, we would expect to see a direct correlation between unconsciousness and the brainwave field's overall strength. In other words, the more unconsciousness there is, the more brainwave activity there is. The less unconsciousness there is, the less brainwave activity there is.

To test this idea, we will explore five kinds of neuroscience experiments. The footnotes for these experiments appear in the Reference section at the back of the book.

I. Anesthesia Experiments

Neuroscientists at Harvard, M.I.T. and Brown recorded the brainwaves of 10 subjects as they were gradually given anesthesia. They noticed that *as the subjects lost consciousness, their brainwave power increased*. (Purdon et al, 2013) (9)

Researchers at M.I.T., Harvard, Boston University and Brown recorded the brainwaves of subjects as they received anesthesia. They found that *loss of consciousness was accompanied by an increase in low beta and high alpha band brainwave power*. (Ching et al, 2010) (10)

Experiment	Consciousness	Brainwave Activity	Footnotes
Anesthesia	C ↓	B ↑	9, 10

II. Fainting Experiment

A team of scientists in Rome tested 63 patients with a history of fainting, and induced unconsciousness using a tilt table. They observed that *loss of consciousness was accompanied by an increase in EEG brainwave amplitude*. When patients regained consciousness, their brainwave amplitude diminished. (Ammirati et al, 1998) (11)

Experiment	Consciousness	Brainwave Activity	Footnotes
Anesthesia	C ↓	B ↑	9, 10
Fainting	C ↓	B ↑	11

III. Exercise Experiments

We can empirically observe that exercise constricts the amount of free conscious awareness a person has available. (This is why guided meditation sessions always begin with relaxation.) It is easy to maintain an expanded state of awareness when comfortably seated in meditation. It is more difficult to maintain the same state while running a 100 yard dash.

If exercise reduces the supply of free conscious awareness, what does it do to brainwave activity?

Researchers at Elon University in North Carolina tested 20 subjects during exercise on a recumbent bicycle. They discovered *brain EEG activity increased during exercise*, and may be related to exercise intensity. Brain EEG activity returned to resting levels quickly after the cessation of exercise. (Bailey et al, 2008) (12)

A team of exercise physiologists in Germany measured 11 subjects during exercises on a treadmill and a stationary bicycle. They found *exercising raised alpha and beta brainwave activity*. (Schneider et al, 2009) (13)

Experiment	Consciousness	Brainwave Activity	Footnotes
Anesthesia	C ↓	B ↑	9, 10
Fainting	C ↓	B ↑	11
Exercise	C ↓	B ↑	12, 13

IV. Meditation Experiments

If brainwave activity increases when people become less conscious, what does it do when people enter into states of higher awareness? A natural starting point for this inquiry would be to study meditation.

Neuroscience researchers at Yale, Columbia and the University of Oregon tested the brainwaves of 12 subjects during meditation. They deliberately restricted their sample to very experienced meditators from a single practice tradition (mindfulness/insight meditation). This approach was intended to reduce heterogeneity in meditation practices. They found *meditation reduced brainwave activity* (Brewer et al, 2011). (1)

Subjective experience of meditative states has also been associated with reduced activity in the brain's default mode network in a study of 32 subjects conducted by researchers at the University of Massachusetts and Stanford (van Lutterveld et al, 2017) (7), as well as in four additional experiments cited by van Lutterveld and Brewer in their 2015 paper (8), including Brewer et al, 2011 (6), Pagnoni et al, 2012 (9), Brewer and Garrison 2013 (10) and Garrison et al, 2015 (11). The opposite effect – distracted awareness with higher default mode network activity – has also been observed by Brewer and Garrison, 2013 (10).

Researchers led by Lisa Miller at Columbia University's Spirituality Mind Body Institute conducted fMRI brain scans of 27 subjects while they recalled personal spiritual experiences. They observed *reduced brain activity* in the left inferior parietal lobule. They also discovered *reduced brain activity* in the medial thalamus and caudate regions when participants recounted spiritual experiences versus neutral or stressful experiences. (27)

Although these results are interesting, they are not conclusive. Meditation comprises so many different kinds of schools, techniques, goals and methods that there is no uniform definition of what meditation is.

Brewer et al states: *"Previous studies have examined individuals using meditation techniques from different traditions (e.g., Tibetan Buddhism, Zen Buddhism, Vipassana, Mindfulness), and employed a wide variety of experimental methods. However, given the methodological differences….no consensus has emerged as to what the neural correlates of meditation are."*

Cognitive Engineering

A team of Italian scientists lead by Barbara Tomosino analyzed 24 neuroimaging studies of meditation and concluded, like Brewer, that the results differed depending upon the type of meditation practiced. (Tomosino et al, 2012). (14)

In short, "meditation" is a slippery word which means many different things to different people. Since there is no agreement on what "meditation" is, or what its goals, methods, or techniques are, there is no consensus on how it affects brainwaves. Consequently, we must look elsewhere to ascertain what happens to brainwave activity when people enter states of higher awareness.

Experiment	Consciousness	Brainwave Activity	Footnotes
Anesthesia	C ↓	B ↑	9, 10
Fainting	C ↓	B ↑	11
Exercise	C ↓	B ↑	12, 13
Mindfulness Meditation	C ↑	B ↓	1

5. Experiments with Psychoactive Compounds

Searching the neuroscience field for laboratory experiments which measure the brainwaves of people in higher states of consciousness also reveals a large body of literature on experiments with psychoactive compounds, which are known to expand consciousness and promote metacognition.

We searched the neuroscience field for laboratory experiments which measured the brainwaves of people in higher states of being. The searches revealed a large body of literature on entheogen experiments.

An *entheogen* ("generating the divine within") is a chemical substance that is ingested to produce a nonordinary state of consciousness for religious or spiritual purposes. Entheogens are used in religious, shamanic, or spiritual contexts that often induce psychological or physiological changes. Entheogens have been used to supplement many diverse practices geared towards achieving transcendence, including meditation, yoga, and prayer. [9]

Entheogens can catalyze intense spiritual experiences, during which users may feel they have come into contact with a greater spiritual or cosmic order. [10]

---

[9] Wikipedia entheogen article
[10] Wikipedia LSD article

Researchers at Johns Hopkins University tested the reactions of 52 subjects to the classic entheogen psilocybin under laboratory conditions. Using a 100-item questionnaire to assess the phenomenological content of altered states of consciousness, they experimentally verified psilocybin's ability to engender *spiritually-significant mystical experiences*. (MacLean et al, 2011) (2)

Neuroscientists at the Imperial College of London and two other UK universities administered psilocybin to 15 volunteers. They observed *profound changes in consciousness* which were accompanied by *significantly decreased brain activity*. They noticed the magnitude of the reduction in brain activity correlated positively with the intensity of the drug's subjective effects. (Carhart-Harris et al, 2012) (3)

A team of neuroscientists in Switzerland tested the effects of psilocybin on 50 volunteers. They found the mind-expanding drug *reduced brainwave current*. They also noticed the intensity levels of psilocybin-induced spiritual experience and insightfulness correlated with desynchronization of brainwaves (which reduces their voltage by wave interference). (Kometer et al, 2015) (4)

Neuroscientists from four universities in the UK tested 15 subjects under psilocybin with functional magnetic resonance imaging (fMRI) and magnetoencephalography (MEG). They discovered that *expanded states of awareness* were accompanied by *large decreases in brainwave oscillatory power* and reduced neural activity. (Muthukumaraswamy et al, 2013) (5)

Researchers from universities in Spain and Austria tested the effects of the mind-expanding psychoactive beverage ayahuasca on 18 subjects. They found ayahuasca *decreased absolute brainwave power* across all frequencies. (Riba et al, 2002) (6)

Neuroscientists from four universities in the UK measured the effects of the psychoactive drug MDMA ("Ecstasy") on 25 volunteers. They found MDMA *reduced brain activity*, and the magnitude of the reductions was highly correlated with the subjective intensity of the drug's mind-expanding effects. (Carhart-Harris et al, 2013) (7)

Neuroscience researchers at the Imperial College of London and three other UK universities summarized the results of several experiments which used different neuroimaging (brainscan) techniques – functional magnetic resonance imaging (fMRI) and magnetoencephalography (MEG) – to understand how psychedelics change brain functions to alter consciousness. They concluded that *consciousness-expanding psychedelics cause brain activity, functional connectivity and oscillatory power to all decrease* in brain regions that are normally highly metabolically active. (Carhart-Harris et al, 2014) (8)

Experiment	Consciousness	Brainwave Activity	Footnotes
Anesthesia	C ↓	B ↑	9, 10
Fainting	C ↓	B ↑	11
Exercise	C ↓	B ↑	12, 13
Mindfulness Meditation	C ↑	B ↓	1
Psychoactive Compounds	C ↑	B ↓	2-8

Cognitive Engineering

VI. Summary

The 5 classes of experiments cited in this chapter reflect the work of 92 scientists on 13 teams in 7 countries using 3 different neuroimaging technologies over a 20 year period. The neuroimaging equipment used was fMRI, MEG, and EEG. The participating countries were Italy, Germany, Switzerland, UK, Spain, Austria, and the USA. The sponsoring universities in Europe were the University of Barcelona and University of Vienna, and in the UK, the Imperial College of London, Kings College, Cardiff University, and the University of Bristol. In the US, the sponsors were Harvard, Yale, Columbia, Brown, M.I.T., Boston University, Johns Hopkins University, University of Oregon, and Elon University. Results are summarized below:

Experiment	Consciousness	Unconsciousness	Brainwave Activity	Footnotes
Anesthesia	C ↓	U ↑	B ↑	9, 10
Fainting	C ↓	U ↑	B ↑	11
Exercise	C ↓	U ↑	B ↑	12, 13
Meditation	C ↑	U ↓	B ↓	1
Psychoactive	C ↑	U ↓	B ↓	2-8

– Table 3 –

Table 3 recapitulates the results of the experiments covered in this chapter, and adds a new column for unconsciousness. Consciousness and unconsciousness are reciprocals per the Law of Conservation. Studying the values in the column for unconsciousness, we notice that it consistently tracks brainwave power. To express this relationship in scientific form, we would write a simple equation:

Formula 4
First Law of Neurodynamics $U_k = B$

where:

U = unconsciousness value (number of applied awareness points)

B = brainwave field electric charge (newtons per coulomb[11])

k = a conversion factor constant (applied awareness points per newton/coulomb)

---

[11] A scientific measurement for electric field strength, often equivalent to volts per meter.

Cognitive Engineering

● Calculation of Brainwave Field Charge

The brain's electric field is generated by current in the brain's electrical circuits – electrically-charged particles flowing through neurons and axons. The strength of the brain's electric field is a function of the current flowing through its neural circuits. This current is calculated by the classical equation:

where:

$$I = \frac{V}{R}$$

I = current in amperes
V = voltage in volts
R = resistance in ohms

This formula is known as Ohm's Law in electrical engineering.

The brain's electric currents can be calculated by a form of Ohm's Law stated as:

where:

$$B = \frac{V}{R}$$

B = brain electric current
V = brainwave field voltage
R = brain neurology resistance

B is the product of the fraction V divided by R.

As the numerator V increases, B will rise. As the numerator V decreases, B will fall.

As the denominator R increases, B will fall. As the denominator R decreases, B will rise.

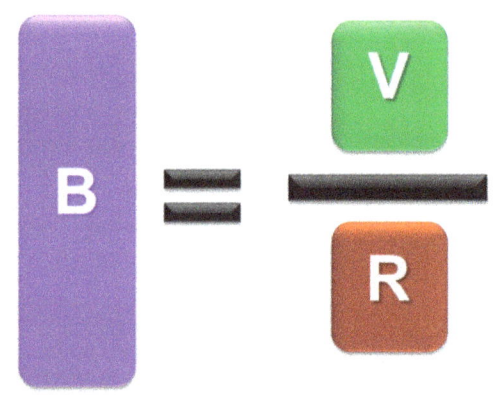

As explained earlier under neurophysics, the neuron acts as a resistor and a capacitor to the electric currents flowing through the brain. The higher the neuron's resistance, the more difficult it is for the current to flow.

Another way of saying this is the neuron's resistance determines its excitability. Highly excitable neurons offer little resistance to incoming electric current. They are very active and fire frequently. Unexcitable neurons present high resistance to incoming current. They are less active and fire infrequently, or not at all. Brain regions exhibit lower activity when fewer neurons are firing, or the ones which are firing go off less frequently, or both.

Cognitive Engineering

The voltage of the brain's electric current and the resistance of the neuron are two independent factors. When either one changes, it does not affect the other. Rather, it affects the total amount of current which is flowing through the brain. This, in turn, determines the strength of the brain's electric field.

Brainwave Field Strength			
Case 1	Case 2	Case 3	Case 4
Higher Voltage	Lower Voltage	Higher Resistance	Lower Resistance
Higher voltage **V** will overcome unchanging neural resistance **R** more easily, causing larger numbers of neurons to fire more quickly. This increases the brain's electric current and its electrical field.	Conversely, a reduction in voltage **V** lessens the brain's electric current and lowers its electrical field.	Higher neural resistance **R** impedes the flow of electric current through the brain, reducing neuron activity. This reduces the brain's electric current which decreases its electrical field.	Conversely, lower neural resistance **R** will increase the flow of current in the brain, which strengthens its electrical field. The smaller denominator **R** increases the value of the fraction.

– Table 4 –

Summary

The brain's electric field is generated by electric current running through the brain's bioelectrical circuits. The brain's electric current can be calculated by a form of Ohm's Law stated as:

where:

$$B = \frac{V}{R}$$

B = brain electric current
V = brainwave field voltage
R = brain neurology resistance

Cognitive Engineering

● **Second Law**

Since B is equivalent to U_k per Formula 4, U_k can be substituted for B in the formula just given, yielding the following equation:

Formula 5
Second Law of Neurodynamics
$$U_k = \frac{V}{R}$$

where:

U = unconsciousness value (number of applied awareness points)

k = a conversion factor constant (applied awareness points per newton/coulomb)

V = voltage of brainwave field

R = resistance of neurology

This law explains the relationship between an embodied conscious being, its level of unconsciousness, and its neurology. When a conscious being incarnates into a body, its unconscious wave component enters into an electrical circuit with the body's nervous system. Since unconsciousness is joined into an electrical circuit with neurology, the interplay between unconsciousness, the being's electromagnetic field, and the body's nervous system will behave according to the laws which govern all electrical circuits.

● **Consciousness Dimension**

This diagram depicts the relationship between brainwave power and consciousness, where:

C = Consciousness
U = Unconsciousness

Per Formula 4, U_k = **B**. Per the Law of Conservation, **U** and **C** are reciprocals. As one increases, the other one decreases.

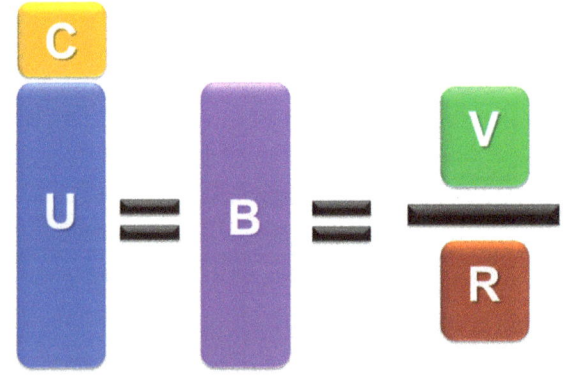

The following diagrams show how changes in brainwave voltage **V** and neural resistance **R** affect consciousness. The results they illustrate are congruent with the experimental findings summarized earlier in Table 4.

Cognitive Engineering

	Consciousness and Brainwave Field Strength	
Case	Diagram	Remarks
Case 1 Higher Voltage V↑ U↑	*As voltage rises, unconsciousness increases.*	Higher voltage **V** can propel greater numbers of electrons through unchanging resistance **R** in the brain's neural pathways.
Case 2 Lower Voltage V↓ U↓	*As voltage declines, unconsciousness diminishes.*	As brainwave voltage **V** declines, it has less power to propel electrons through unchanging resistance **R** in the brain's neural networks. This reduces brainwave activity, and lowers the brain's unconscious life force power consumption **U**.
Case 3 Higher Resistance R↑ U↓	*As resistance increases, unconsciousness diminishes.*	Higher neural resistance **R** impedes unchanging brainwave voltage **V** from pushing electrons through the brain's neural networks. This lowers brainwave activity, reducing the brain's unconscious life force power consumption **U**. Lower unconsciousness **U** exposes greater consciousness **C** per the Law of Conservation.
Case 4 Lower Resistance R↓ U↑	*As resistance falls, unconsciousness rises.*	As neural resistance **R** diminishes, unchanging voltage **V** can propel greater numbers of electrons through the brain's neural pathways. This raises brainwave activity, increasing the brain's unconscious life force power consumption **U**. Higher unconsciousness **U** comes at the expense of consciousness **C** per the Law of Conservation.

– Table 4 –

Cognitive Engineering

● Resistance

The resistance cases just described (cases 3 and 4) dovetail with what science knows about how psychoactive substances work in the brain.

As shown in the first diagram, neurons transmit electrons to one another over the junction between neurons called the synapse.

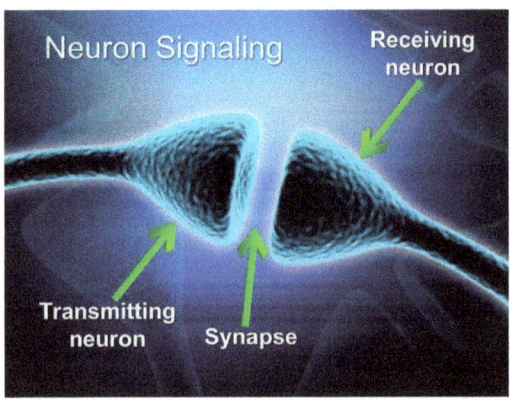

As depicted in the second diagram, electrons enter the receiving neuron through doorways called "receptor sites."

Neurotransmitter inhibitors, such as those found in psychoactive drugs (entheogens), block some of these receptor sites, thereby raising the neuron's resistance to incoming electrons.

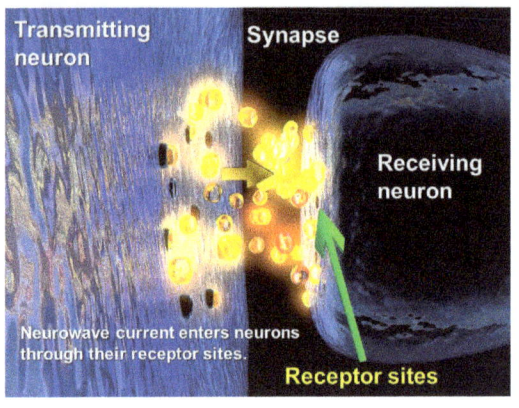

Blocking either neurotransmitter receptors or the synthesis of neurotransmitters impedes the flow of electrically-charged ions from one neuron to another. The property of electrical conductors which impedes the flow of charged particles within the conductor is called *resistance*.

As discussed earlier under "Propagation Level 4 – Neurowaves," neurons collect incoming charged ions into a cellular reservoir which, when filled, causes the neuron to fire. If the flow of ions between neurons is impeded, the neurons become less active (i.e., fire more slowly).

When unconscious brainwave activity diminishes, the living awareness that was intertwined with

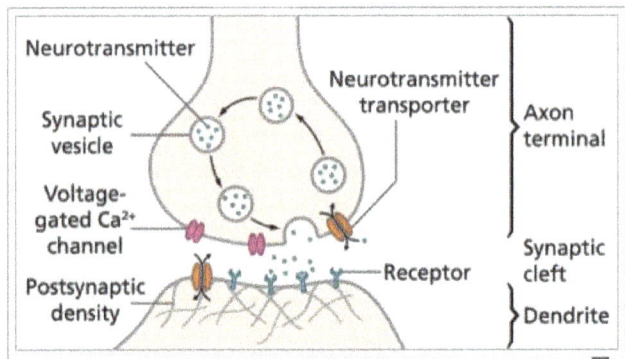

Illustration of the major elements of neurotransmission. Depending on its method of action, a psychoactive substance may block the receptors on the post-synaptic neuron (dendrite), or block reuptake or affect neurotransmitter synthesis in the pre-synaptic neuron (axon).

Source: Wikipedia
Psychoactive drugs article

brainwaves is freed, and it coagulates into a larger pool of reflective self-awareness, yielding higher states of being.

Formula 5
Second Law of Neurodynamics

$$U_k = \frac{V}{R}$$

where:

U = **Unconsciousness.** Number of awareness points applied to the unconscious memory wave field (which condenses into the electromagnetic brainwave field). Each of these applied awareness points contributes force to a memory wave. All awareness points are accounted for, yielding a closed system. Awareness cannot be created or destroyed.

V = **Voltage.** Sum of all neurowave amplitudes. All brainwave amplitude originates from the power of applied awareness points which generate memory waves. The memory wave field includes all active memory history, as well as the memory templates for:

(a) all active memory history

(b) the memory templates for:
   1. interpreting incoming perceptions
   2. activating outgoing behavior
   3. supporting language and other cognitive functions

R = **Resistance.** Neurology resistance. Regulates how easily neurowaves move through the system. When R increases, it reduces the number of applied awareness points flowing through neurology in the form of memory waves that are condensed into neurowaves. When R decreases, their number increases.

Note: This law only describes incarnated behavior. Between lifetimes, the being has no neurology, so the resistance R in the formula is zero. Division by zero is not allowed, so when the being is not incarnated, different laws apply.

Cognitive Engineering

● Summary

The principle of Propagation (Corollary 3) and the two Laws of Neurodynamics mathematically describe how the electrical interplay between consciousness and neurology creates and animates embodied life.

An incarnated conscious being's unconscious memory wave component enters into a tightly-coupled electrical circuit with its body's neurology. The being's unconsciousness supplies applied awareness points which generate memory waves that power the brainwave field.

The brain is like an appliance. It draws current in amperage. Current and charge are proportional. Unconsciousness is proportional to the overall electric current in the brain.

The more memory waves there are, the greater number of applied awareness points there are. The higher each memory wave's amplitude is, the more applied awareness points it draws. More brainwaves equal more unconsciousness and fewer brainwaves equal less unconsciousness.

The strength of the memory field emanating from a conscious being is proportional to the number of applied awareness points enclosed by the being. (This phenomenon can be viewed as a higher octave of Maxwell's First Law, which states that the electric field leaving a volume is proportional to the charge inside.)

Since unconsciousness is joined into an electrical circuit with neurology, the dynamic interplay between unconsciousness, the being's electromagnetic field, and the body's nervous system behaves according to the laws which govern all electrical circuits.

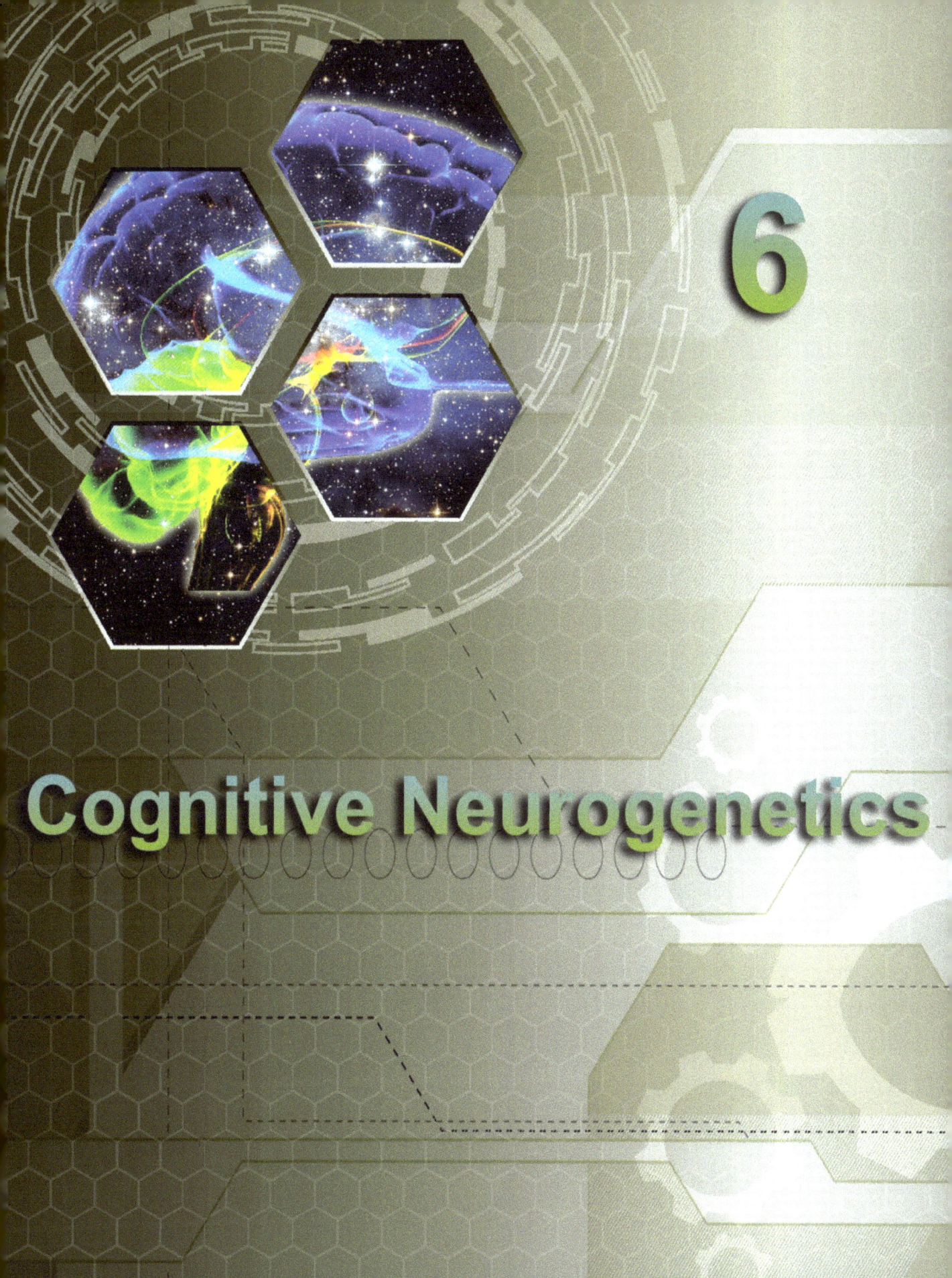

Cognitive Neurogenetics

6

Neurogenetics Design Influences

Cognitive physics informs the design of cognitive neurogenetics.

Cognitive Mechanics

- The periodic table indicates humanity occupies the lower 20% of the awareness scale.

- The law of conservation stipulates that if a conscious being is 20% conscious, they are 80% unconscious. If we want more conscious awareness, this law informs us we can retrieve it from unconscious awareness.

- The propagation corollary shows how unconscious awareness manifests itself as memory waves, which condense into brainwaves. Unconsciousness equals memory waves which equal brainwaves.

Meta-Memory Science

- Conscious free awareness becomes unconscious applied awareness by animating active memory records which emit memory waves.

- The memory power corollary explains how high amplitude memories, such as those containing trauma, require the greatest number of awareness points to animate.

- High amplitude memories also take the longest time to deactivate. Thus, strong past life memories are likely to remain in force, while low amplitude memories go inert.

Cognitive Neurophysics

- Neurophysics codifies the mind/matter interface between the conscious being and the physical neurology it inhabits.

- It descirbes the layer-by-layer connection between active unconscious memories and electrically-charged particles moving through the brain's neural networks.

- It shows how the conscious being's non-physical memories are intertwined with the brain's magnetic fields and electrochemistry.

Cognitive Neurodynamics

- The neurodynamics experiments demonstrate that reducing brain activity can restore free awareness to the conscious being.

- Neurodynamics explains how an incarnated conscious being's unconscious memory waves are in a tightly coupled electrical circuit with its body's neurology.

- If the body's neurology causes brainwaves to collapse, then memory waves that are intertwined with brainwaves also collapse. This releases applied (unconscious) awareness points that were animating the memory waves back into pure conscious free awareness.

Genetic Engineering

● The New Genie

A new laboratory tool called CRISPR has sparked a revolution in genetic engineering. CRISPR[12] is an advanced genome editing system that allows scientists to manipulate the genes of any living creature with astonishing ease. This breakthrough technology could increase longevity, fortify health, and even create designer humans.

CRISPR enables scientists to edit DNA as if they were programming a computer. CRISPR technology yields a tremendous improvement in DNA editing speed, ease, cost, accuracy and precision. It is being hailed as a monumental landmark in the history of biological research

Genetic engineering is a foundational science—on par with physics, chemistry, and electronics. With powerful tools like CRISPR, transhumanism concepts we might think of as science fiction today can be converted into mainstream reality overnight. The genie is out of the bottle.

The development of CRISPR is being compared to the invention of the transistor, which spawned the Information Age and ushered in the Internet. With this new technology, humanity is now entering a Genetic Age.

Gene therapy can be used in hundreds of ways to improve the human body, including longevity, health, wellness, mental acuity, intelligence, beauty, stamina, strength, immunity and resiliency.

It is technically feasible to genetically engineer higher awareness into DNA with CRISPR. Today's conscious young people need to know that this technology offers exciting possibilities for their lives. People in their twenties and thirties today could conceivably spend most of their lives in higher states of being.

---

[12] short for Clustered Regularly Interspaced Short Palindromic Repeats

Gene Editing

Genome editing science is advancing so rapidly that any snapshot taken of it will be out-of-date in six months. Accordingly, the following discussion presents a high-level overview of technology trends in the genome editing field rather than a rigorous treatment of current genome editing techniques.

Humanity's history with the Industrial Revolution and the Information Age suggests that the Genetic Age we are now entering will witness rapid and profound technological advances. As this new age dawns, there are currently several ways to edit human genes:

1. *Gene knockout* is a genetic technique in which one of a person's genes is made inoperative.

2. *Gene knockdown* is a process which reduces the expression of one of a person's genes by a fixed percentage, such as 30%.

In the future, there will likely be hundreds of gene editing methods available which can adjust individual gene expression with a fine precision (e.g., anywhere from 1% to 100%).

Future advances in genetic engineering will yield dozens of different methods we can barely imagine today to change and optimize the expression of genetic traits.

CRISPR

An editing method known as the CRISPR–Cas9 technique delivers gene edits to cells by inserting the genetic material into a hollowed out virus, which serves as a shipping container to transport the edits to cells. Cells absorb the virus and the genetic material splices itself into the cell's DNA.

CRISPR is an acronym for "clustered regularly interspaced short palindromic repeats," a description of the genetic basis of this method. Cas9 is the name of an enzyme which is able to reprogram the DNA in human cells.

CRISPR represents a series of DNA sequences discovered in bacterial microbes for defending against attacking viruses. These microbes manufacture thousands of forms of CRISPR. If they can all be harnessed, we may see future advances in genetic engineering that we can barely imagine today.

Viruses use bacteria as miniature factories to replicate themselves. Viral microbes introduce their genetic material into bacteria, tricking the bacteria's cellular machinery into using it as blueprints to produce viral proteins. Viruses can use human cells in the same way. CRISPR exploits this mechanism by removing a virus's genetic material and substituting therapeutic DNA. The re-engineered virus can then introduce the therapeutic DNA into human cells.

Future forms of genome editing may employ advances even more revolutionary than CRISPR. Microbes have evolved many different defense mechanisms against viruses, some of which are only now being discovered. Scientists have recently found that microbes can use another group of proteins, called Argonautes, to dismantle viral DNA. Chinese researchers have successfully used Argonaute proteins to edit DNA in human cells. These proteins could yield yet another powerful gene-editing tool.

● Gene Regulation

Almost all the cells in the human body contain the exact same DNA, but each cell type has a different set of active genes. Through a process called gene regulation, a cell controls which genes, out of the many genes in its chromosomes, are "expressed" or activated to make functional proteins. Different patterns of gene expression cause the various cell types to produce different sets of proteins, customizing each cell type to perform its uniquely-specialized function in the body.

● Why Gene Editing Works In Neurons

Unlike other human cells, neurons generally do not reproduce (except in a few areas of the brain such as the hippocampus). Most neurons in the brain and nervous system last for the lifetime of the individual. How, then, can genetic engineering change these cells?

For their long life, neurons must rely on the human body's self-regulation systems which identify worn-out proteins and replace them with new ones. Human cells experience a constant turnover of nearly every one of their components. Every cell is involved is involved in a continual process of breakdown and re-growth that is essential to life. During this maintenance and repair activity, cellular proteins are broken down into their component amino acids, which are then re-used as building blocks for cellular renewal.

Replacement proteins are made using the cell's DNA templates. If the cell's DNA or RNA has been altered, it will affect the manufacture of replacement proteins. For example, if editing has turned off the gene controlling a type of cellular function, then the replacement proteins for that function will not be made. The half-life of proteins in human cells can be up to several days long, so changes from genetic editing are normally fully expressed within a week. [13]

Special delivery methods are required to transport genetic edits through the brain's capillaries into neuron cells, and these are well understood by contemporary science.

● RNA Editing

Researchers have also found that an ordinary mouth bacterium makes a form of CRISPR that can edit RNA — the molecular messenger which turns DNA's genetic information into cellular proteins. This discovery, known as CRISPR interference, further enlarges the library of tools available for genetic engineering.

In neurons, RNA translates genetic blueprints into proteins which are used to build every part of the cell, including its receptor sites. Imagine a neuron which has 3 receptor types: A, B and C. Suppose we silence the RNA messenger which translates the gene for the B receptor type into the proteins for building it. The next time the cell rebuilds itself, it will only have A and C receptors.

Except for DNA, all contents of a cell including RNA are recycled every few days, so editing RNA produces a temporary result which lasts for a several days. RNA edits could theoretically enable cognitive enhancement candidates to "test drive" new states and acclimatize themselves to living in higher awareness before proceeding ahead to permanent genetic upgrades.

RNA editing can also provide genetic R&D programs with an engineering test bed for quickly exploring different editing strategies to optimize technical approaches.

---

[13] DNA is an exception to the half-life rule, lasting for the lifetime of the individual. DNA owes its long life to dedicated repair mechanisms that patch up damage. For cellular proteins however, no comparable repair mechanisms are known to exist. They must be recycled and rebuilt.

Cognitive Engineering

To modify DNA to support higher states of awareness, genetic engineers have to understand how a conscious being interacts with its neurology. As the Cognitive Neurophysics chapter explains, causing brainwaves to collapse liberates the applied awareness points which animate them. Hence, the strategy for raising free, unbounded conscious awareness is to attenuate certain kinds of brainwave activity.

The mission of neurogenetics is to select unconscious brainwaves which are not essential for the individual's normal functioning in life, and then dampen these waves to restore higher levels of conscious awareness.

Gene therapy can be deployed to raise consciousness in many different ways by reducing the excitability of certain types of neurons in the brain. Scientific experiments can be conducted to isolate neural pathways and regions whose activity can be safely attenuated to raise mental acuity, emotional well-being and cognitive abilities without compromising the individual's overall functioning in the world.

- Disrupting Painful Memory Activation

A conscious being can retrieve memories into its conscious or unconscious awareness. This text refers to conscious memory retrieval as *recall* and unconscious memory retrieval as *activation*.

Applied awareness is stuck in active memory records. Free awareness can be restored by draining applied awareness out of memory.

The Memory Power Corollary states that the number of applied awareness points required to animate a memory is proportional to the memory's amplitude. Highly-charged emotional, painful memories have the greatest amplitude of any type of memory. Hence, they soak up the largest number of applied awareness points when animated.

Cognitive Engineering

The strength of emotions and sensations determines their brainwave signal power, so memories charged with pain and strong negative emotions typically exhibit high amplitudes. Consequently, the more emotionally-charged negative memories are active in a human being's unconscious mind, the fewer free awareness points the being has.

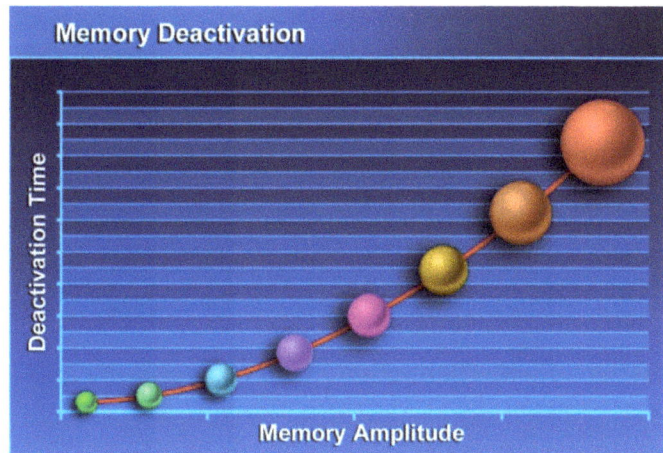

This graph predicts average memory deactivation time in relation to memory amplitude.

High amplitude memories require longer to deactivate than low-amplitude memories. For example, traumatic memories generally take longer to fade than minor injuries, as their emotional and sensory amplitude is higher.

If high amplitude memories have the greatest persistence, then emotionally-charged memories such as traumatic incidents are more likely to be carried forward through multiple lifetimes than milder memories which dissipate more quickly.

This suggests that a major portion of unconsciousness could be comprised of a conscious being's active, emotionally-charged, painful past life memories, and that these unconscious memories could be encasing large regions of awareness.

Gene therapy to impede the unconscious activation and retrieval of negative emotional memories can prevent them from siphoning off awareness points from the conscious being. Since these kinds of emotional memories tend to have high amplitude, the potential savings in awareness can be substantial.

Vast reserves of human cognitive potential are buried in unconscious memory waves which can be attenuated by re-engineering certain parts of the central nervous system. This chapter presents a roadmap for future genetic engineering programs for unlocking this latent cognitive potential to enhance consciousness. It discusses four regions which appear to hold promise as candidates for cognitive engineering.

● Raising Intelligence

Lower brainwave current not only results in higher awareness, it also raises IQ.

Scientists at the Ruhr University in Bochum, Germany have discovered that higher IQ individuals have fewer dendrites in their brains. A team of researchers led by Dr. Erhan Genc analyzed the brains of 259 subjects using neurite orientation dispersion and density imaging, which enabled them to measure the amount of dendrites in the cerebral cortex. All participants completed IQ tests which were correlated with their neuroimages. (24)

The results showed that the more intelligent a person is, the fewer dendrite connections there are between the neurons in their cerebral cortex. Using a database from the Human Connectome Project, Dr. Genc's team confirmed these results in a second sample of 500 individuals.

Receptors are located on dendrites. Fewer dendrites means fewer receptors. Fewer receptors yields higher resistance. Higher resistance reduces unconsciousness, thereby increasing consciousness.

$$U = V / R \qquad R\uparrow = U\downarrow \qquad U\downarrow = C\uparrow$$

Dr. Genc's report also cites other studies which have shown the brains of highly intelligent people demonstrate less neuronal activity during an IQ test than the brains of average individuals. One such study, conducted by Dr. Richard Haier at the University of California, found *significantly lower brain activity* in subjects during an abstract reasoning test, as indicated by cortical metabolic rates measured with positron emission tomography. (25)

Neuronal activity is measured in voltage. Lower voltage reduces unconsciousness, thereby increasing consciousness.

$$U = V / R \qquad V\downarrow = U\downarrow \qquad U\downarrow = C\uparrow$$

The conscious being is the seat of human intelligence (not the brain). The more consciousness is present, the more intelligence is available for the individual to draw upon.

Cognitive Engineering

● Lowering Excitability

Neurons build their receptor sites from their DNA blueprints. A neuron's genetic blueprints can be modified to build fewer receptor sites.

● Receptor Choice

Today's genome editing is a nascent technology in an primitive stage. Yet even at this early point, gene edits can be focused on specific cell types, such as neurons, through the use of precision targeting navigational aids called *guides* and *vectors*.

As the technology evolves, it will likely develop the ability to isolate neuron sub-types in specific brain regions. In the beginning, though, a simple, direct approach of targeting the most promising neuron receptor type will yield the best results.

There are 50 different types of neuron receptors, but one receptor outshines all the rest in terms of safety, potency and efficacy. The serotonin 2A receptor (5-HT2AR) is a prime candidate for genetic cognitive neuroengineering since it is very well known and has been researched in hundreds of studies. Neuroscience experiments associate down-regulating this receptor with reduced brainwave power and expanded states of awareness. All classic entheogens including psilocybin and LSD chemically bind to the serotonin 2A receptor in neurons. Experiments have shown a strong positive correlation between an entheogen's affinity for the 5-HT2A receptor and its psychoactive potency.

Entheogens temporarily make neurons less excitable by chemically blocking the electrical conductivity of the serotonin 2A receptor sites in their dendrites. Genetic engineering can permanently lower the excitability of these neurons by physically reducing their population of serotonin 2A receptor sites. This can be accomplished by down-regulating the expression of gene HTR2A on chromosome 13, which codes the expression of the serotonin 2A receptor. This gene is an ideal candidate for editing as it is chemically dissimilar to its neighbors on the chromosome and minimally polymorphic, meaning there are fewer varieties of the gene. Less receptors means fewer pipes for electric current to flow into the neuron.

Brainwaves traveling down neural pathways through neurons with the 5-HT2A receptor release greater awareness when their power is attenuated. Genetically dampening the excitability of neurons having this receptor will elevate consciousness.

LSD and Neurons

1. LSD crystals lodge in receptors: When LSD molecules brush against receptor sites, the crystals get stuck, much like billions of tiny kites getting stuck in billions of tiny trees. (31) The crystals block incoming ions from entering the receptors.

2. The crystals stay in the receptors: When LSD latches onto a neuron's serotonin receptor, the LSD molecule is locked into place because part of the receptor folds over the drug molecule like a lid and it stays put. Dr. Bryan Roth, a professor of pharmacology at University of North Carolina, and his team of scientists at UNC captured crystallography images showing the atomic structure of an LSD molecule bound to a human serotonin receptor. They discovered that the LSD molecule was wedged into the receptor's binding pocket, and part of the receptor protein had folded in over the LSD like a lid, sealing the drug inside. (32) (33) (34)

3. LSD crystals stop neuron signaling: When Dr. Roth's lab at UNC exposed live neurons in a Petri dish to microdose-sized amounts of LSD, they found even tiny doses of LSD affected the receptors' signaling. Specifically, they found LSD reduced signaling at the 5-HT2A receptor. (32) (33) (34)

According to the University of Bristol in England, the body 'mistakes' LSD for serotonin, and shoots it across the synaptic cleft. LSD has a higher affinity for 5-HT receptors than serotonin, thus the presence of LSD prevents serotonin from sending neural messages in the brain. Once the LSD molecule is bound to the receptor proteins, the message is not carried any further. (35) Receptors with LSD crystals attached are blocked from receiving electrically-charged neurotransmitter ions.

4. LSD crystals do not activate neurons: Many psychoactive compounds are technically receptor activators ("agonists") but their concentrations in the body are too low to cause neurons to fire. Although LSD is an agonist, and it binds to the 5-HT2A receptor as if to activate it, there are so many neurons in the brain – and the concentration of LSD is so low – that there are not enough LSD molecules attached to the average dendrite to actually activate it and cause it to fire. Most 5-HT2A receptors are affected at too low affinity to be sufficiently activated by the average brain concentration of LSD (approximately 10–20 nanoMolars). (36) (37)

Cognitive Engineering

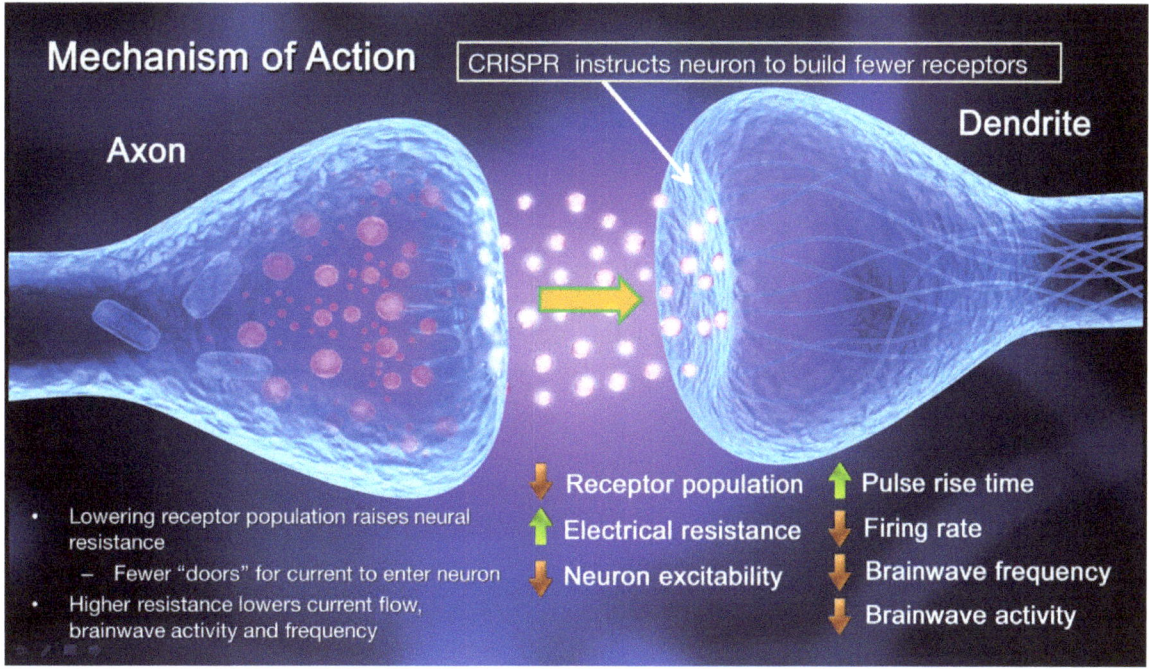

- Neuron Excitability

A neuron's receptor sites serve as doorways which receive the flow of electrically-charged ions into the neuron. A neuron will fill its cellular reservoir with incoming charged ions more quickly if it has a larger number of receptor sites.

Increasing the number of a neuron's receptor sites adds more channels for incoming ions to flow into, similar to adding more lanes to a freeway. This gives the neuron lower electrical resistance, which makes it more easily excitable.

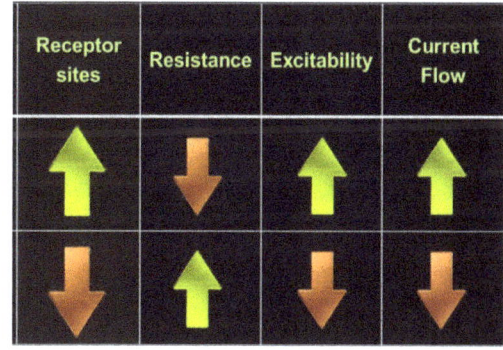

Conversely, decreasing a neuron's receptor population reduces the number of pipes for incoming ions to flow into, like closing lanes on a freeway. This raises the neuron's electrical resistance, making it harder to excite.

A neuron's resistance can be modified by changing its number of receptor sites. Reducing a neuron's number of receptor sites by removing its 5-HT2A receptors decreases the number of doorways or pipes for electrically-charged ions to flow through, thereby increasing the neuron's resistance. This decelerates the flow of

173

electrons from one neuron to another. Raising a neuron's resistance lowers its conductivity. Less-conductive neurons have a lower capacity for carrying the flow of electrical current in the brain.

● Brainwave Attenuation

A moving electrical current generates an electromagnetic wave (per Ampere's Law). Flowing electrons in the brain generate brainwaves. When the flowing electrons slow down, so does brainwave activity. Less-conductive, less-excitable neurons require more time to fill their cellular reservoirs with enough electrically-charged ions to cause them to fire. Hence, they fire less frequently. Lower neuron activity reduces brainwave activity.

Numerous scientific studies have conclusively demonstrated reduced brainwave activity is correlated with higher states of awareness, concentration, focus, mental acuity and cognitive ability. Accordingly, attenuating the subject's brainwave activity will yield a cognitive enhancement.

● Current

A psychoactive drug's potency is highly correlated with its affinity for the serotonin 2A receptor, but serotonin is only one of seven major kinds of receptors. All those other receptors are wide open. How can a drug make such a large change in consciousness by causing only a minor overall reduction in resistance? The answer is that a small change in resistance causes a large change in current per Ohm's Law, as illustrated in the following chart.

Cognitive Engineering

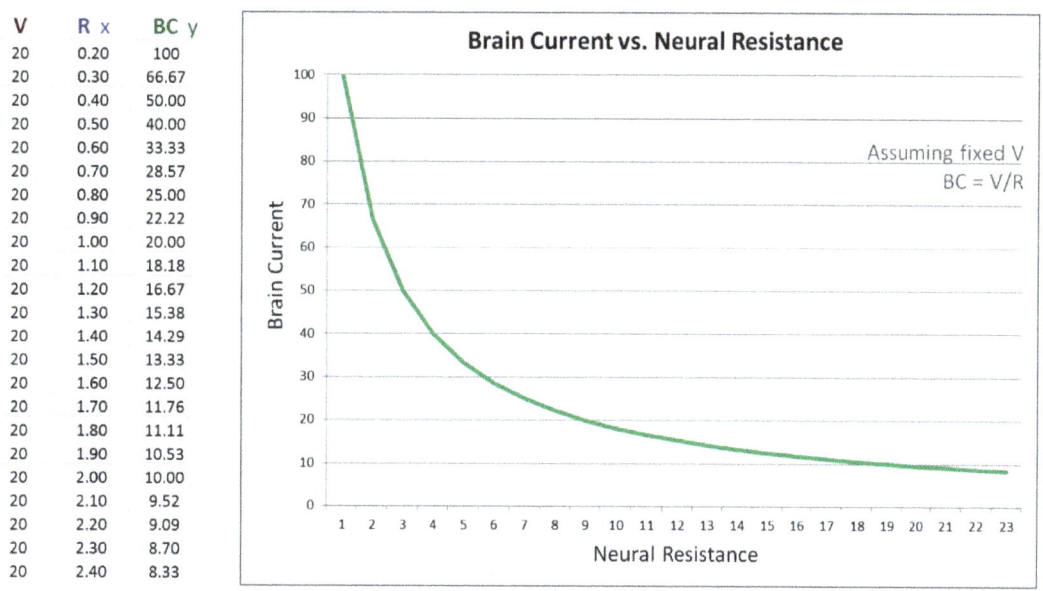

V	R x	BC y
20	0.20	100
20	0.30	66.67
20	0.40	50.00
20	0.50	40.00
20	0.60	33.33
20	0.70	28.57
20	0.80	25.00
20	0.90	22.22
20	1.00	20.00
20	1.10	18.18
20	1.20	16.67
20	1.30	15.38
20	1.40	14.29
20	1.50	13.33
20	1.60	12.50
20	1.70	11.76
20	1.80	11.11
20	1.90	10.53
20	2.00	10.00
20	2.10	9.52
20	2.20	9.09
20	2.30	8.70
20	2.40	8.33

● Voltage

Some of the experiments cited report reductions in brainwave voltage accompanying expanded states of consciousness. Per Ohm's Law, increasing resistance lowers brainwave current (BC=V/R) but not voltage. How do we reduce voltage?

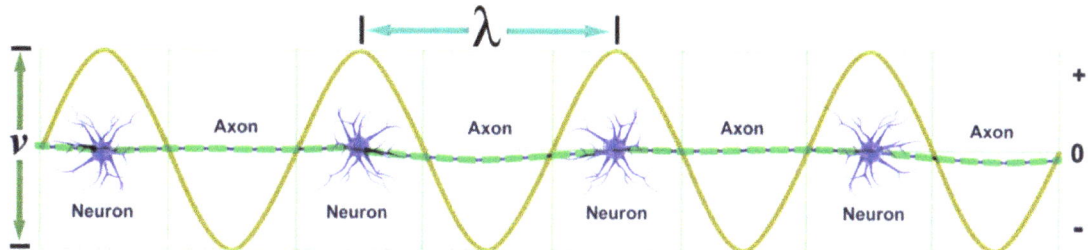

The answer is that raising a neuron's resistance lowers its conductivity, which disrupts the frequency signatures of brainwave signals traveling through the neuron (represented here by the Greek letter lamda).

Cognitive Engineering

This frequency shift causes brainwaves that were once synchronized to become desynchronized, resulting in wave interference when the peaks and valleys of desynchronized brainwaves intersect. When waves that were once synchronized partially cancel each other out, it reduces brainwave field voltage.

Also, altering a brainwave's frequency causes it to lose its structural integrity and information content. For example, if you were to disrupt the frequency signature of a voice recording during playback, it would become meaningless noise. This disruption in turn causes the associated memory waves to lose their information and release awareness points.

● Summary

Consciousness level varies inversely with brainwave power. Unconscious brainwave activity can be attenuated by lowering neuron excitability. Genome editing can dampen neuron excitability by reducing the population of receptor sites.

● Posterior cingulate cortex

The neuroscience experiments discussed in the previous chapter provide additional evidence for the posterior cingulate cortex's role in the retrieval of painful memories. We cited a number of experiments where subjects experienced reduced brainwave activity and expanded spiritual awareness in response to psychoactive substances known as entheogens.

As stated earlier, an entheogen is a chemical substance that is ingested to produce a nonordinary state of consciousness for religious or spiritual purposes. As discussed in the Cognitive Neurodynamics chapter, researchers at Johns Hopkins University tested the reactions of 52 subjects to the classic entheogen psilocybin. Under laboratory conditions, they experimentally verified psilocybin's ability to engender spiritually-significant mystical experiences. (MacLean et al, 2011) (2)

Psilocybin, LSD and all classic entheogens chemically bind to the serotonin 2A receptor (5-HT2AR) in neurons in the brain. As noted earlier, there is a strong positive correlation between an entheogen's affinity for the 5-HT2AR receptor and its psychoactive potency. In humans, the distribution of these serotonin 2A receptors is *densest in the posterior cingulate cortex* (Carhart-Harris et al., 2012). (3)

Impeding the activation channels for painful memories by interfering with neural pathways in the posterior cingulate cortex is part of the mechanism by which entheogens release awareness from the past into the present.

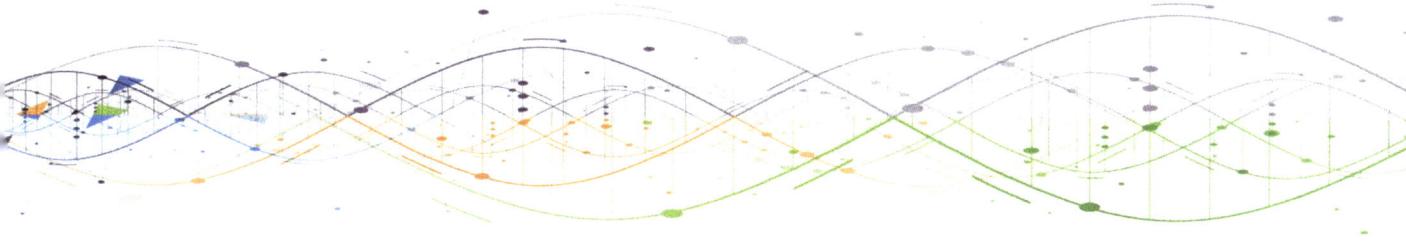

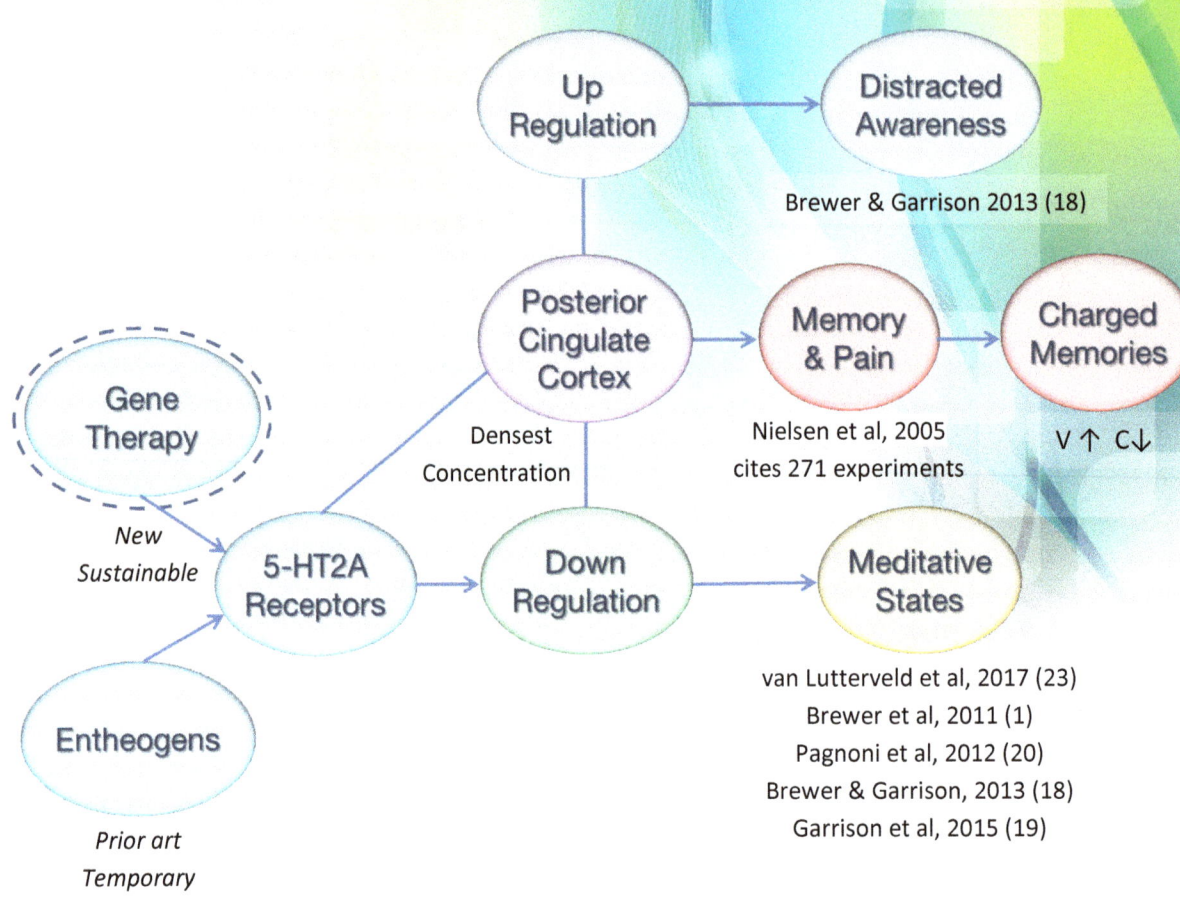

Reduced activity in the posterior cingulate cortex has also been associated with subjective experiences of meditative states in a study of 32 subjects conducted by researchers at the University of Massachusetts and Stanford (van Lutterveld et al, 2017) (23), as well as in four additional experiments cited by van Lutterveld and Brewer in their 2015 paper (17), including Brewer et al, 2011 (1), Pagnoni et al, 2012 (20), Brewer and Garrison 2013 (18) and Garrison et al, 2015 (19). The opposite effect–distracted awareness with higher PCC activity–has also been observed by Brewer and Garrison, 2013 (18).

Since a human being's unconscious awareness is largely dedicated to animating unresolved traumatic memories of the past, we would expect to see a noticeable recovery of free awareness when the neural channels for retrieving painful memories were disrupted or blocked. Accordingly, a prime strategy for neurogenetics is to dampen the brain's painful memory retrieval circuits.

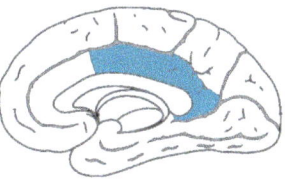

To achieve this, neurogenetics researchers must first identify the brain's neural circuits responsible for painful memory retrieval. A region of the brain thought by many scientists to manage painful memory retrieval is the posterior cingulate cortex.

Over the years, opinions have varied as to the functions of the posterior cingulate cortex. To develop a consensus, a team of Danish scientists led by F.A. Nielsen performed a meta-analyses of a large number of neuroscience experiments to determine the main functions associated with this region. They found the prominent roles for the posterior cingulate cortex are episodic *memory retrieval* and *pain*. (Nielsen et al, 2005) (16)

Compared to older sciences like chemistry and physics, which are hundreds of years old, neuroscience is still a very young field, and it is continuously evolving. There is still no universal agreement among scientists on the functions of many areas of the brain, including the posterior cingulate cortex. Nevertheless, since Nielsen's team analyzed a rather large body of evidence – 271 experiments in all – their findings are interesting and may be useful to cognitive engineers.

● Conclusions

This body of evidence suggests that gene therapies to dampen the 5-HT2AR receptors densely expressed in the posterior cingulate cortex could lower the brainwave field's electric charge (V↓) and yield long-lasting gains in free conscious awareness (C↑).

Although memory and pain are the primary functions identified for the PCC, numerous other studies have implicated the PCC in a several other functions, such as mind wandering, craving, and judgmental thinking. Interestingly, these functions represent cognitive behavior often associated with diminished free awareness (C↓) – a by-product of painful memory activation.

Researchers have linked PCC activity to cognitive processes of being "attached to" or "caught up" in one's experience, such as craving, and more subtle experiences of getting "caught up," such as identifying with or being attached to attributes or labels of ourselves (Brewer and Garrison, 2013).

This role for the PCC is also supported by data showing that the PCC *decreases* in activity when we are not caught up in experience, whether being focused on a task or meditating. Specifically, the subjective experiences of "undistracted awareness" and "effortless doing" correspond with PCC *deactivation*, and "distracted awareness" and "controlling" corresponded with PCC *activation*.

Craving, perhaps one of the most obvious experiences of being caught up in experience, is described clinically and experimentally in terms of desire, urge, want, and need (Tiffany and Wray, 2012) and it has been associated with PCC activity in smoking and drug addiction.

Clearly, this is a region ripe with possibilities for cognitive engineering. Researchers must bear in mind, however, that humanity is only just beginning to grasp neuroscience, and there are strict limits to our knowledge.

For example, although wise teachers have cautioned humanity for centuries about the pitfalls of judgmental thinking, part of the PCC's evaluative activity occurs during social cognitive processing, which can include the resolution of moral dilemmas involving issues of benevolence, compassion, social justice and fairness.

Cognizant the limits of our current knowledge, scientists must tread slowly and carefully. Fifteen sub-regions of the posterior cingulate cortex have been identified. The dorsal (upper) regions have been associated with cognitive processing, while the ventral (lower) regions are involved with memory. Additional research can identify the best engineering strategies for harvesting the cognitive potential of this area with the fewest side effects.

Cognitive Engineering

● Amygdala

Research indicates the amygdala is extremely active during emotional situations, and acts with the hippocampus and prefrontal cortex in the encoding of emotional memories.

The amygdala learns and stores information about emotional events, so it participates in emotional memory. Emotional memory is an implicit or unconscious form of memory and contrasts with explicit or declarative memory mediated by the hippocampus.

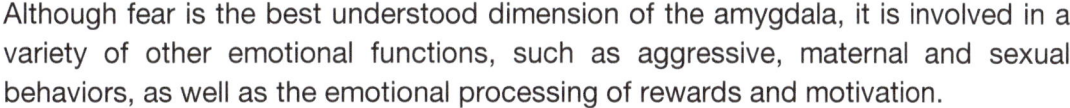

The flow of information through the amygdala's neural circuits is regulated by four kinds of neurotransmitters (dopamine, norepinephrine, epinephrine and serotonin) which influence how its excitatory and inhibitory neurons interact.

Although fear is the best understood dimension of the amygdala, it is involved in a variety of other emotional functions, such as aggressive, maternal and sexual behaviors, as well as the emotional processing of rewards and motivation.

In addition to its role in emotion and unconscious emotional memory, the amygdala is also involved in the regulation of cognitive functions, such as attention, perception, and explicit memory. These functions are modulated by the amygdala's processing of the emotional significance of events. Outputs of the amygdala lead to the release of chemicals in the brain that alter cognitive processing.

● Research Directions

Neuroimaging experiments can be conducted to identify which types of neurons in the amygdala are most active during stimulation of fearful or traumatic memories. Gene therapies can then be designed to lower the excitability of these specific neurons by reducing the expression of receptor sites in their dendrites. Fewer receptor sites means higher electrical resistance, as there are fewer pipes which incoming current can flow through.

Cognitive Engineering

● Constraints

The amygdala is involved in the encoding and retrieval of both positive and negative emotional memories. Researchers must ensure that any gene editing of the amygdala does not affect the encoding of positive memories.

For instance, a team of 13 Japanese neuroscientists found that dopamine D1 receptors located in the dendrites of amygdala neurons play a major role in potentiating the amygdala response. (Takahashi et al, 2010) (15) This research needs to be refined to identify which types of neurons are active in the amygdala during the encoding of emotionally pleasurable memories. If neurons can be isolated in the amygdala which encode emotionally traumatic memories, but not positive ones, they would represent candidates for gene therapies to boost awareness.

● Summary

Our early ancestors lived in a dangerous world filled with predatory foes. The amygdala, *flight-or-fight* reactions, and stimulus-response reflexes represent reptilian brain animal survival neurology.

Civilization has made our environment much safer now. Perhaps the time has come for the human race to step forward and leave the animal kingdom. We now possess the technology to begin attenuating reptilian brain functions which are no longer necessary, and trade them up for higher conscious awareness.

● Nociceptors

When the human body has a physical issue, it must alert its owner to the problem. However, the human nervous system has a reputation for delivering considerably more pain than is necessary to bring problems to its owner's attention. Some would say this represents a design flaw. In any event, humanity now has the power to correct it.

As humankind progresses further into the 21st Century, biotechnology has become sufficiently advanced for scientists to design pain out of human bodies with genetic engineering and alleviate the source of trauma. In the future, there may be no need for your nervous system to alert you to physical issues directly. Theoretically, it could call your smart phone.

Editing gene SCN9A segments W897X, I767X, and S459X will block the Nav 1.7 sodium ion channel used by nociceptors (pain nerves) for conveying pain signals to the brain. There are no discernible physical abnormalities associated with the change, and discriminative touch is not affected.

Admittedly, modifying gene SCN9A is an unsophisticated approach. Removing pain nerves entirely from the nervous system design is a more elegant solution. But editing this gene will work. Just knocking down SCN9A could wipe out 70% of human physical suffering.

Dampening nociceptors can restore awareness from unconscious memories back to the conscious being in present time.

Whenever someone experiences physical pain, the perception activates painful memory records in their unconscious. This siphons off awareness points from consciousness into the activated memory records. High amplitude memory records soak up the most awareness and persist for the longest time.

Unconscious painful memories come into activation and drop out of activation without the person's conscious awareness. The next time the person experiences pain, it may activate different memories.

People living painfree lives will not re-activate traumatic memories, nor will they refresh or reinforce any of the active painful memories they brought with them into the lifetime. As a result, the lifetime is one of memory deactivation and not activation. Eventually, over decades, the gradual deactivation of painful memories will build up a noticeable dividend of free awareness which is restored to the conscious being.

In addition, as pain and fear are drained from the human life experience, the amygdala will be activated less and less frequently. Increasing disuse over the course of generations could contribute to a reduction in the amygdala's volume.

🟢 Very long term memory

Have you ever considered or examined the baggage you brought with you into this lifetime? Human beings usually do not incarnate with a clean slate. The "ghosts" of our old personas come trailing in with us, carrying their own memories, beliefs, behaviors, stuck points of view, unresolved issues and traumas.

The old identities who are still stuck in unresolved trauma and survival patterns continue to draw the most energy from us, draining our life force and awareness. These ghosts of our old personas are also working overtime to perpetuate the past.

Our past life memories are electromagnetic wave signatures, which we generate continuously, and carry with us from life to life. When we are incarnated, they manifest themselves as brainwaves in the long term memory regions of the brain's temporal lobes and other areas. When we decarnate, they come with us.

India's great wisdom traditions assert that conscious beings can have hundreds of past lives. If this is true, then a person's current life memories represent less than one percent of their total memory. Accordingly, unconsciousness is thought to be primarily composed of active past life memories. All of these memories are being animated by our own personal awareness, which could be put to better use serving us in present time.

Neuroimaging experiments can be conducted to pinpoint areas in the temporal lobes, hippocampus and posterior cingulate cortex which light up during past life regressions. Scientists can then analyze these regions to isolate their specific neuron receptor types, and create gene therapies to lower their excitability. This approach can theoretically yield expanded conscious awareness without any side effects.

Cognitive Engineering

● Exploration

Navigating uncharted territory always carries an element of risk of the unknown. From the earliest oceanic explorers right up through today's astronauts, humanity has repeatedly ventured into the unexplored, overcoming its challenges and discovering its treasures.

Genetic engineering today is at a nascent stage of development akin to where computers were in 1960. The earliest mainframes were limited to processing accounting applications by sequentially reading reels of magnetic tape. The scale of potential improvements in genetic engineering is on the same order of magnitude as the distance the computer industry has travelled from 1960 to today. Gene therapies for elevating consciousness are positioned to ride the power curve of these forthcoming advances in the decades ahead.

Although the CRISPR editing tool is amazingly powerful, it is only the beginning of a long procession of advances to come. Edits done today will seem primitive to future generations when people look back on them twenty or thirty years from now. Although the number of gene editing and regulation options are limited today, there will be hundreds of options available tomorrow.

NASA did not shoot for the Moon straight out of the gate. They reached their objective in stages. The Mercury program established the feasibility of sending a man into space and returning him safely to the Earth. The Gemini program sent two men into orbit for longer periods of time, and verified their ability to perform docking and space walks. The third program, Apollo, finally sent three men to the Moon.

Similarly, cognitive engineering will achieve its objective in stages. It must first prove it can genetically raise a person's free awareness to a higher plane, and then bring them back safely to their normal level. Next, it will create sustainable higher states of awareness which people can function in for long periods of time. Thirdly, it will develop the ability to produce finely-tuned states of being which enhance specific attributes of consciousness and yield different meditative qualities.

Limitations and Opportunities

The most permanent solution for optimizing a conscious being's progress to advanced states is to remove all traces of traumatic, emotionally-charged memories from their unconscious mind. Any of this baggage the person has been carrying around with them for untold lifetimes is something they ultimately want to jettison.

Cleaning house of painful past life memory energy signatures requires cognitive therapy. Therapeutic protocols for addressing this issue are provided by a new cognitive science, trans-life psychology, described in the book *Future Life Design*.

Gene therapy is not going to clean house. It will merely block the expression of active traumatic memories while the being is incarnated. The results can last for a lifetime, but when the being decarnates, its negatively-charged memories can be reactivated by new painful experiences if it reincarnates into a body with pain nerves.

Should the human race choose to widely embrace the technologies in this book, however, then future generations of humans may no longer experience physical pain. Anyone reincarnating into one of these generations would not encounter painful stimuli which could reactivate traumatic memories.

Nevertheless, the most comprehensive solution for achieving advanced states of being which are stable over many lifetimes is still cognitive therapy to discharge negative memories.

Future Directions

This chapter has identified four gene therapy targets, but there are many more waiting to be discovered. Scientific research and experiments can verify, refine and tune the gene targets and gene therapy methodologies discussed here.

This book's purpose is to illuminate an overall direction and strategy for achieving results. Its insights provide a roadmap for neuroscience research experiments and genetic engineering programs to formulate effective gene therapies for enriching human awareness and cognitive capital. Hundreds of questions and issues will arise in the course of this research and engineering. This is normal for any high-tech R&D program. A book like this cannot begin to answer every question or address every issue – but top-notch teams of scientists and engineers can. They will be the heroes in the story of humanity's evolution to higher states of consciousness in the 21st Century.

7

Cognitive Ecosystem

● Overview

Psychologists agree the two most important emotions in human experience are love and fear. These emotions color a person's mental geography and shape the decisions they make which affect their career, relationships, family, and financial status.

Living under the influence of fear limits a person's quality of life. Fear affects the entire life experience, because it affects how people make the day-to-day decisions which add up to determine the quality of their lives. Without this limiting influence, human beings can be free to realize their highest potentials in life, and rise to greater levels of achievement.

● Unconscious Fear and Memory

The fear which influences human behavior is primarily unconscious, not conscious. It draws its energy largely from painful memories which are active in the person's unconscious mind below their level of awareness.

Physical pain reduces consciousness to lower levels of awareness, like fear and anger. After a painful experience, a conscious being's memory continues to hold the emotional force of the experience. The emotional energy is locked in the memory the same way voltage is stored in a battery.

The stored emotional energy of any given memory can be dormant, like an unattached battery, or it can be active, like a battery which is coupled to a circuit. When a traumatic memory is reactivated later on, the "battery" is connected to an individual's unconscious, and the stored emotions of fear or anger are transported into the person's present experience of life. Fear is the most problematic, because it is the most insidious.

From the standpoint of behavior, it does not matter how much potential emotional voltage there is, only how much of it is actively connected to the lifetime. There can be hundreds of these memories actively in force concurrently, contracted over the course of many lifetimes. The sum total emotional voltage of all these memories actively in force manifests itself as attitudes, emotions and behaviors which are fear-based. Eventually, the emotional energy in the painful memory subsides, but, as explained earlier, it can take quite a long time, depending on the severity.

Cognitive Engineering

● Fear-Based Behavior

As a traumatic memory ages, its force gradually dissipates, and it manifests itself in milder forms as attitudes and behaviors which are fear-based. These subtle layers of fear often masquerade as "normal" behavior, such as:

doubt	*worry*	*risk-aversion*
narrow thinking	*reluctance to change*	*anxiety*
uncertainty	*excessive caution*	*hesitation*
distrust	*lack of confidence*	*concern*

These attitudes are what hold people back. Their opposites – love, embracing change, trying new things, and going forward into the unknown – are the positive energies which generate success in life.

Fortune favors the bold. Success requires taking risks. People under the influence of fear are averse to risk, and they do not take the decisive actions which success requires. Reluctant to change, they remain stuck in old behavior patterns.

● Cognitive Engineering Benefits

Cognitive engineering increases an individual's supply of free conscious awareness, which is its primary benefit. The engineering strategy which generates this dividend in conscious awareness also attenuates fear, yielding a substantial secondary benefit.

As explained in the preceding chapter, neuroscience experiments have pinpointed the brain region responsible for managing unconscious painful memories. CRISPR can be used to lower the excitability of neurons in this region, reducing their capacity to transport painful memory brainwaves. This attenuates high-power brainwaves, which liberates significant awareness from unconscious memories. Since these brainwaves represent painful memories, this strategy also reduces fear.

● Developing New Behaviors

The astute application of cognitive engineering can produce genetic upgrades which raise a person's conscious awareness and diminish their underlying fear. To realize the full measure of benefits, upgraded individuals can receive psychological counseling to translate their new cognitive assets into more expansive life strategies. As explained in this chapter, counseling can help them to discover, learn, practice and reinforce new behaviors which lead to success and contribute to personal growth.

Cognitive Engineering

The socially-responsible implementation of neurogenetics demands that gene therapies be buttressed by comprehensive education, counseling and community support programs to prepare individuals for genetic enhancement and support them in achieving their goals.

As illustrated in Fig. 14, professionally-certified counselors screen applicants to determine suitable candidates for gene therapy.

Approved candidates receive education to prepare them for genetic enhancement and share best practices for managing life in an elevated state of consciousness.

Graduates from this program receive counseling to help them harvest the greatest benefit from expanded awareness and remove any blocks or limitations which might interfere with their progress. Participants who successfully complete their education and counseling programs become eligible for genetic upgrades.

Living in higher consciousness will be transformative for many people. Graduates may experience positive shifts in personality, outlook, values, openness, relationships, professional interests, and many other areas. Genetically-enhanced individuals attend continuing education programs which help them understand and navigate these changes to realize the greatest benefit for themselves, their families, their communities, and humanity. They also join other enhanced individuals in community networking groups, and receive additional counseling to help them achieve mastery in managing their lives in higher consciousness.

When individuals elect to receive additional neurogenetics edits, the whole process repeats itself.

Service Process

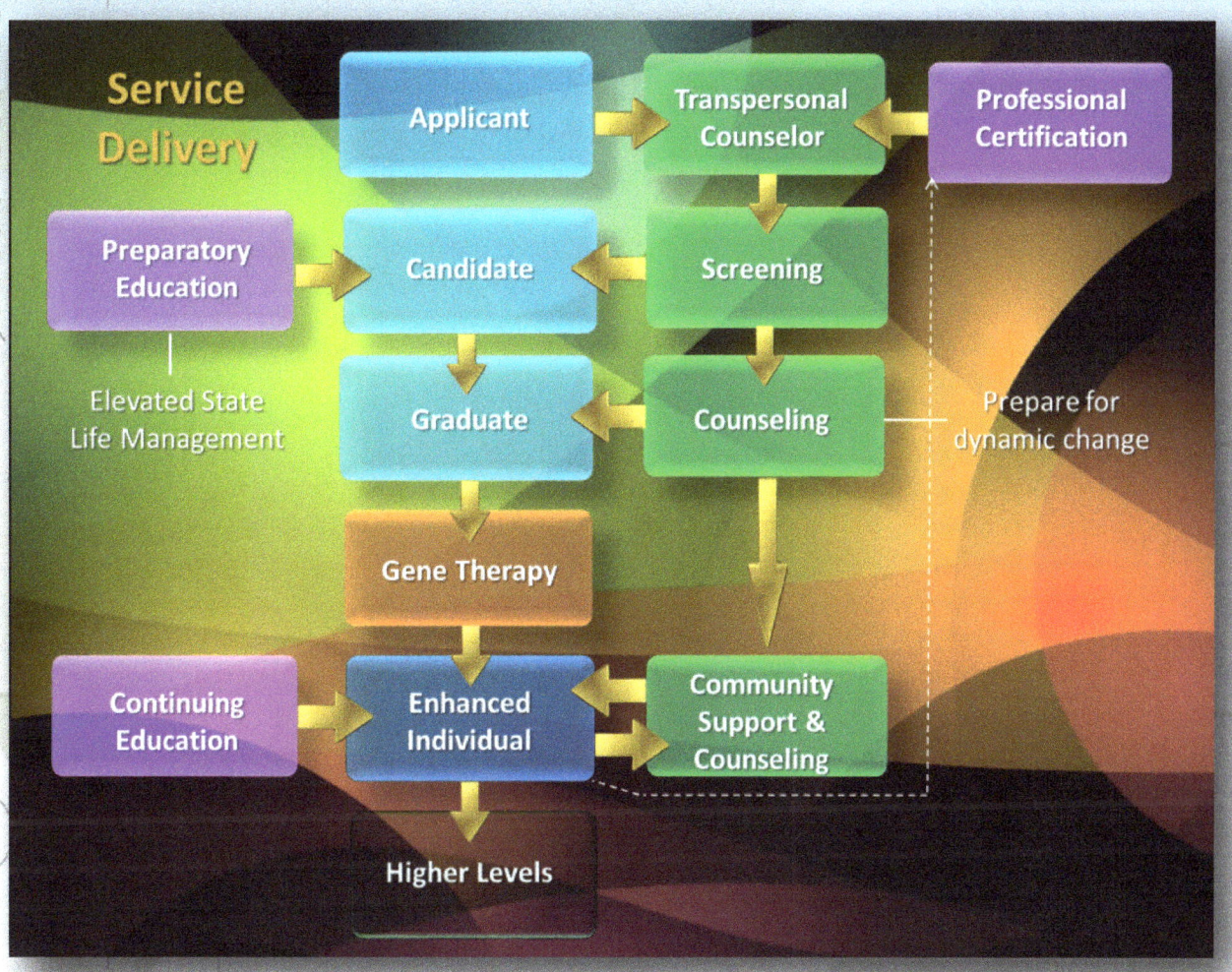

– Fig. 14 –

The results of each upgrade are scientifically verified with brainwave measuring equipment and psychological tests to quantify awareness gains and improvements in cognitive functions.

01 Education

Many learning modalities can be utilized which enable genetically-enhanced individuals to actualize the full benefits of elevated awareness. This subject has a worldwide reach and its delivery model must accommodate a broad range of constituents in different cultures and socio-economic classes. Accordingly, its learning communities will deliver a variety of programs to engage a wide spectrum of students at different levels of education and proficiency.

A curriculum of courses can be developed for delivering a robust education program to genetically-enhanced individuals for maximizing the benefits of elevated awareness in their lives.

Each course can teach higher consciousness management skills from a different angle. Some courses may touch upon many facets of the subject, while others zero in on just a few.

Giving students a diverse menu of programs which they can choose from based on their individual interests, preferences and learning styles yields an eclectic program which serves each student's unique needs.

An education program which deploys a combination of diverse modalities and approaches can produce a synergy and deliver a holistic learning experience. This is a better method than offering a "point solution" that only addresses a single aspect, or a rigid structure that puts everyone on the same track.

Leading universities can deliver a rich tapestry of modalities and regimens to their student bodies. Students can customize their academic programs of study based on their own individual needs, styles and interests.

Rather than specifying one path that everyone must follow, cognitive education should offer an open system architecture which supports a variety of learning modalities to serve a diverse student body who have a wide variety of learning styles and unique individual needs.

Cognitive Engineering

This open systems framework for the study of genetically-enhanced consciousness can create a collaborative environment where leading educators and innovators from a broad spectrum of disciplines can build a multi-dimensional curriculum.

● Preparatory Education

Preparatory education is initially provided in the candidate's local community by a certified counselor. The counselor will introduce the candidate to genetic graduates in the community to hear their stories and answer questions. The counselor will also bring the candidate's family into their office to educate and prepare them for the upcoming changes and answer their questions.

A second module of preparatory education is delivered centrally at a university. Students will receive an overview of the full menu of genome edits to learn about the available choices and how the levels interrelate. They can also study a textbook like this one to grasp the science behind the journey they are embarking upon.

Educators will prepare students for a major positive life change and set expectations for the events to follow. Students will also meet genetic graduates on campus who will share their experiences and answer questions.

Classroom instruction will explain theory and practice for pre-therapy counseling, genetic enhancement, community support groups, ongoing counseling and continuing education.

The curriculum will also include specific modules which are different for each edit.

1. The SCN9A edit syllabus will teach students how to manage painfree living.
2. The HT2AR edit curriculum will train students how to manage living in higher consciousness.
3. The amygdala edit classes will focus on positive emotions and transcending survivalism.
4. The long term memory edit courses will cover reducing the influence of past lives on consciousness.

● Continuing Education

Genetically-enhanced individuals will return to universities for continuing education programs to refine their expanded consciousness management skills. Students will review and discuss their life experiences with others in classroom environments and receive expert guidance on improving their life performance.

Continuing education students will also participate in psychometric testing to establish scientific measurements of genetic engineering results. Test instruments can include EEG monitoring as well as psychological assessments including personality and consciousness questionnaires. Results will be reported in scientific journals.

02 Counseling

Many different kinds of therapeutic modalities can be utilized to enable genetically-enhanced individuals actualize the full benefits of elevated awareness.

The creation of counseling methodologies for genetic enhancement program participants can leverage over half a century of progress in human potential development, drawing from diverse disciplines including transpersonal psychology, integral psychology, cognitive psychology, mindfulness psychology and trans-life psychology.

Professional psychologists in these areas can structure creative approaches for helping people reap the full measure of benefits from genetically-enhanced states of being. These counseling programs are envisioned in four parts:

1. Screening. Counselors will interview genetic program applicants and screen out individuals who have a history of mental illness or neurological disorders, life situations which may impede their progress, dependency on certain prescription medications, and other factors.

2. Preparatory Counseling. Gene therapy candidates who have successfully graduated from preparatory education will receive counseling to condition them for positive change. This therapy will focus on expanding the subject's model of the world, opening them up to change, removing obstacles and limitations, and making sure their life is in shape to support a major positive improvement. It will also help subjects to gain a completely positive mental outlook on their upcoming change, including the development of attitudes, emotions, beliefs, assumptions, values, expectations, decisions, perceptions, thoughts and feelings.

3. Responsive Counseling. Living in higher consciousness will be transformative for many people. Genetically-enhanced individuals may experience positive shifts in personality, outlook, values, openness, human relationships, professional interests, and many other areas. Ongoing counseling can help them understand and navigate these changes to create the greatest benefit for themselves, their families, their communities, and humanity.

4. Proactive Counseling. The shift into greater self-awareness can be leveraged as a powerful new enabler for personal growth. Stable, elevated awareness-of-awareness serves as a constant reminder of one's larger identity, creating an opening for change. Proactive counseling programs can capitalize on this opportunity and lift people out of living in personality-based identities (largely fixated in the past) into more expansive, awareness-centered identities empowered in the present.

Professional counselors can guide enhanced individuals to explore all the dimensions of their new states of being and help them access the full range of cognitive resources which are now available to them. Like polishing a rough new diamond, proactive counseling can bring out the myriad facets of an upgraded individual's new, expanded self.

Liberated awareness yields not only elevated consciousness – it also supplies new raw materials and building blocks for personal growth, including creativity, compassion, focus, clarity, energy, emotional balance and well-being. Skilled teachers and counselors can help cognitively-enhanced people discover how to channel their newfound freedoms into fruitful avenues of personal growth, including character development, self-image and the setting and achieving of goals. Well-designed education and counseling programs can serve as springboards to propel individuals into more empowered versions of themselves.

Preparatory, responsive and proactive counseling programs can all use a common set of techniques and protocols to help people realize maximum benefits throughout the process of cognitive enhancement.

Cognitive Engineering

Leading professionals from various human potential development disciplines can formulate many different elements for these counseling methodologies. The example shown here depicts a four-step program to help students design more conscious lives. It includes building a foundation, developing a blueprint, taking specific action, and harvesting the results.

Foundation
Students are prepared for dynamic change by examining 14 core areas to ensure they have a solid foundation for consciousness expansion. Key areas such as beliefs, emotions and attitudes are reviewed to identify and resolve any hidden blocks to progress.

Blueprint
Students develop a blueprint for their new lives including their purpose, goal, intention and vision. The blueprint serves as a basis for creative visualization.

Action
Students re-examine their priorities and develop future-focused practices to prepare themselves for more effective performance in life.

Result
The program multiplies the student's freedom and range of choices, opening a gateway to a brighter future and a channel for personal growth.

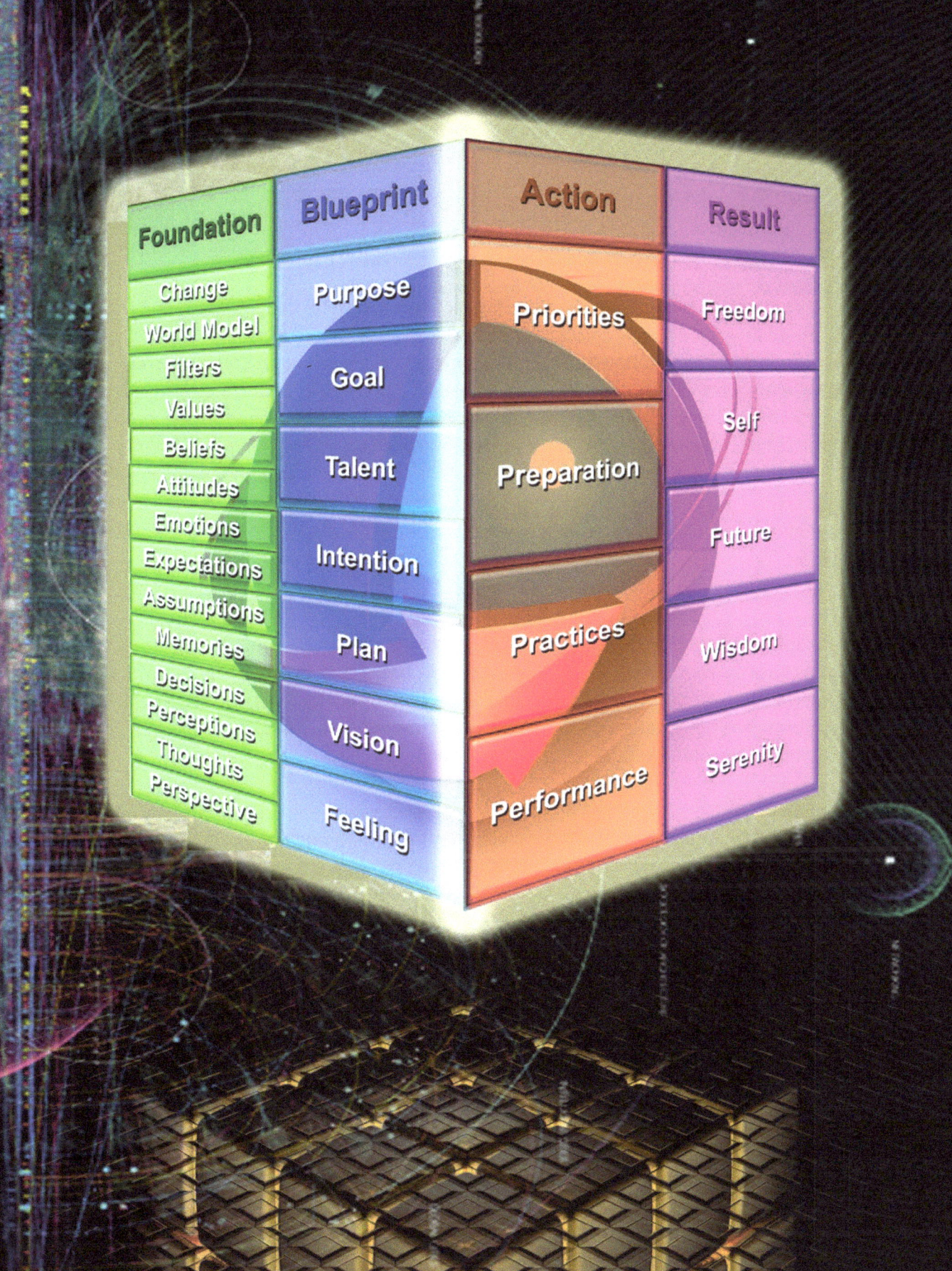

Cognitive Engineering

03 Genetics

Access to gene therapy resources is granted to candidates who have successfully completed preparatory education and counseling. The first step in gene therapy is to establish a psychometric baseline for each candidate using brainwave monitoring equipment and psychological tests. Once this is complete, the candidate can proceed into gene therapy.

Following gene therapy, the battery of psychometric tests is again administered to each enhanced individual to quantify their progress over their baseline measurements. Improvements are compared to the overall goals set for each edit. When satisfactory results have been verified, the edit is certified.

EEG testing will quantify brainwave profile gains in default consciousness and meditation. Psychological tests will assess positive shifts in personality, values, outlook, openness, alertness, mental acuity, compassion, creativity, harmony in human relationships, well-being and happiness.

Genetically-enhanced individuals whose baseline level of consciousness now equals their previous level in meditation will receive the benefits of meditation all day long. Hence, the expected results include the cornucopia of benefits that have been proven for meditation by hundreds of studies over the years…multiplied by an order of magnitude.

04 Community

Counselors will host networking and support groups for genetically-enhanced individuals in their local communities. Through these forums, graduates can share their successes and challenges in the practicalities of living life in higher awareness. They can also discuss their experiences in navigating the transformative changes which arise out of greater personal freedom and an expanded perspective. Friendships may develop among participants who share common ground. Group members can also serve as ambassadors of genetically-enhanced consciousness to their local communities, sharing their stories with families, friends, neighbors and local media.

Summary

Cognitive physics comprises five new interconnected sciences which work together to deliver a revolutionary strategy for raising human conscious awareness through genetic engineering.

1. *Cognitive Mechanics* applies scientific methods to the study of awareness and how it interacts with matter. Its axioms and formulas reveal cardinal laws governing the relationship between awareness and the physical universe.

2. *Macro Memory Science* maps the topography and behavior of a conscious being's unconscious mind, and presents a scientific explanation of the mechanics of non-physical memory.

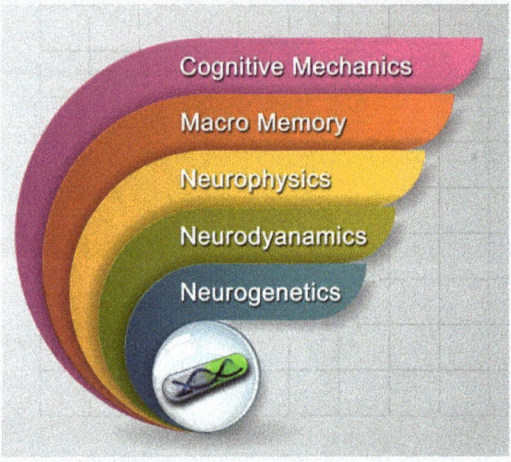

3. *Cognitive Neurophysics* illuminates the relationship between the conscious being and the neurology it inhabits, and reveals the spiritual origins of brainwaves.

4. *Cognitive Neurodynamics* describes the dynamic interplay between consciousness and neurology in scientific equations.

5. *Cognitive Neurogenetics* defines a strategy for genetically optimizing human neurology to support sustainable higher states of being.

Cognitive physics provides a flexible, open systems architecture. Many improvements and refinements can be made to its formulas, laws, methods and techniques. The contributions of future generations of researchers will help humankind to realize the full promise of this exciting new branch of science.

| **Edit 4**
 VLT Memory | Personal sovereignty and self-determination |

| **Edit 3**
 Amygdala | Freedom from unwanted emotions |

| **Edit 2**
 Nociceptors | Freedom from physical suffering |

| **Edit 1**
 Posterior Cingluate Cortex | Restore conscious awareness |

| **Foundation** | Preparatory education programs conveying best practices, skills and techniques |

The edits discussed in the Neurogenetics chapter are summarized above. Although the edits are numbered, they can be performed in any order. Individuals will select the edits they desire based on their own personal priorities, goals and interests.

The edits can each be layered into several separate component levels to ensure a gradual and manageable rate of change.

Although this chart only has four types of edits, future researchers will undoubtedly add many more.

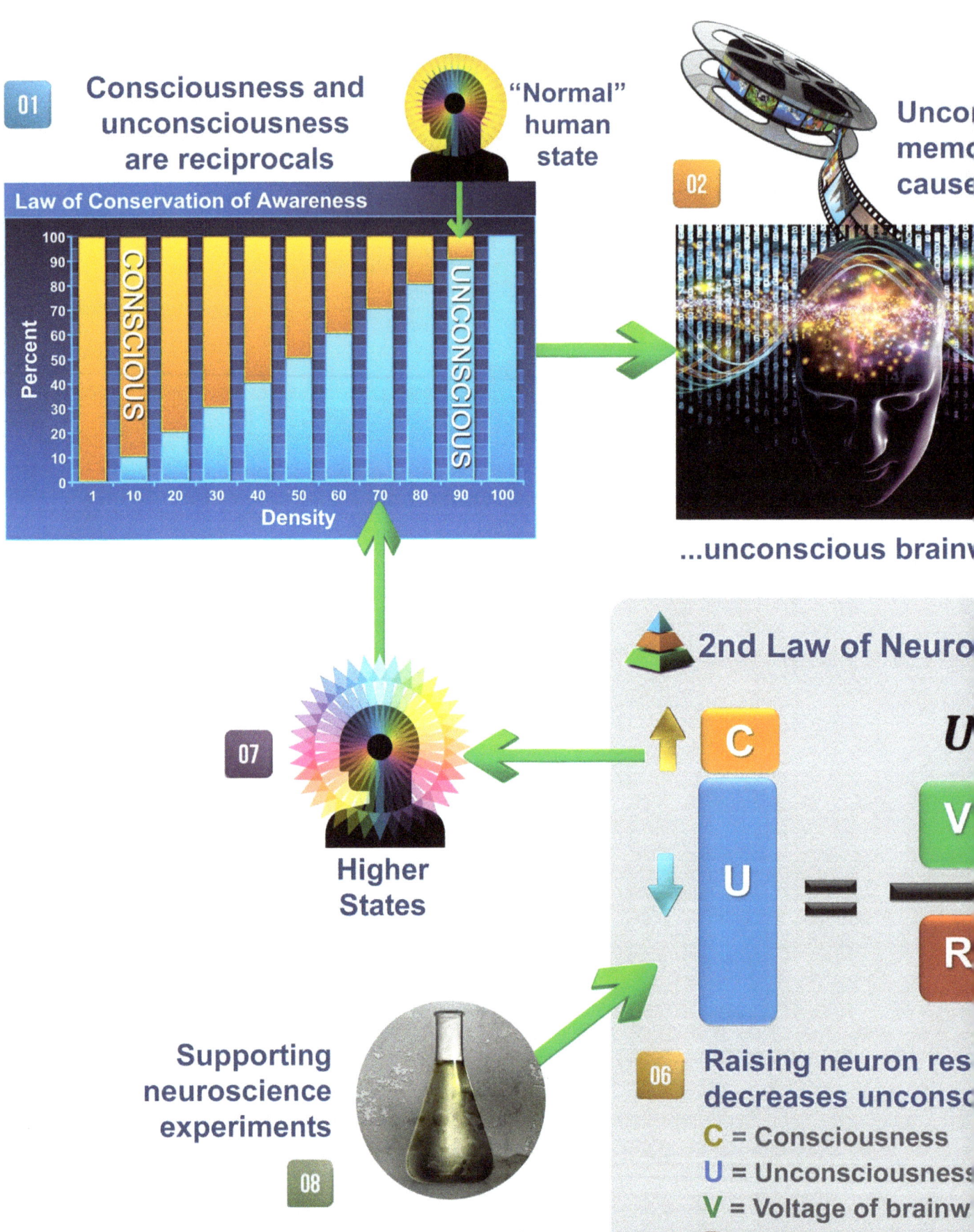

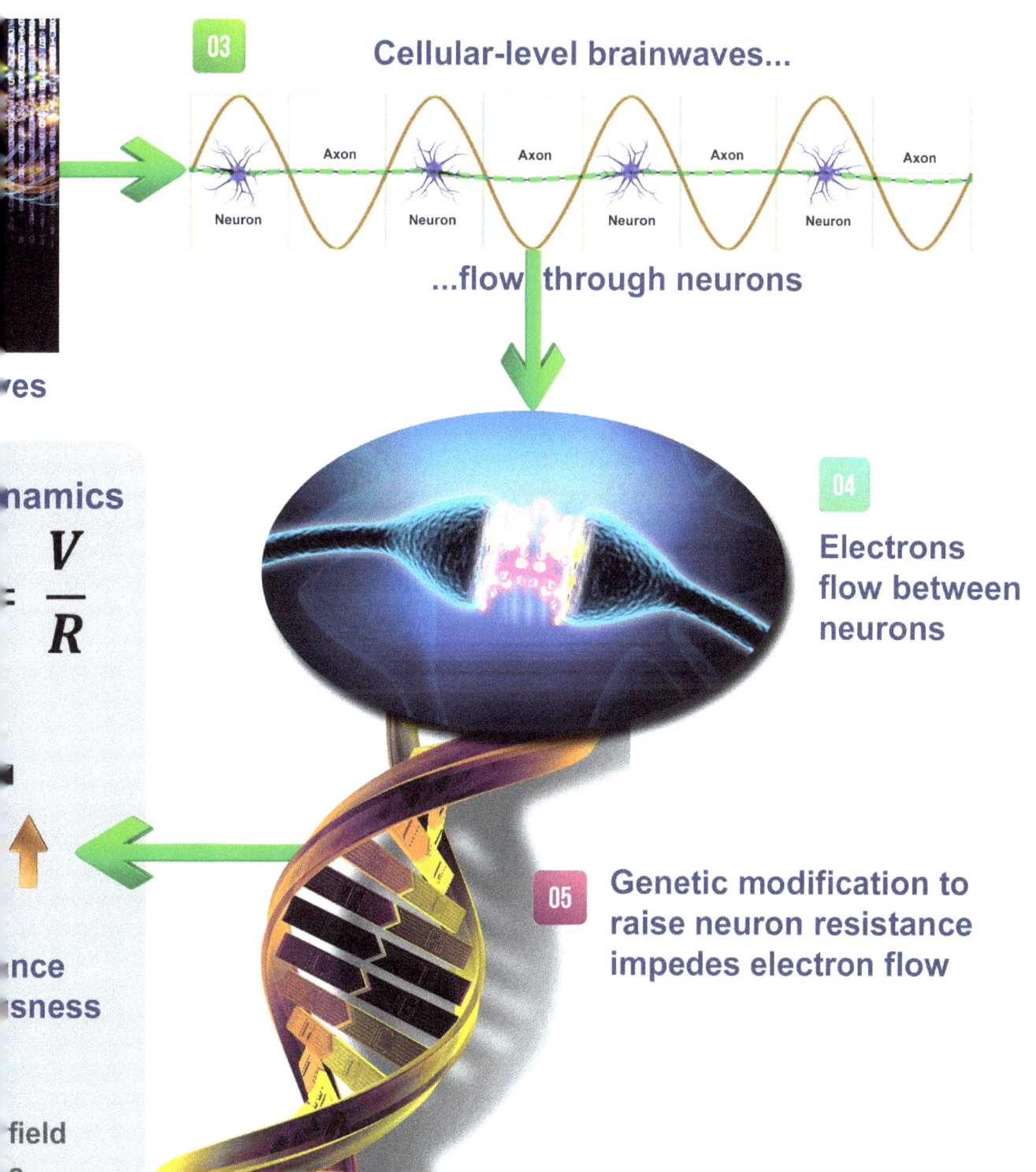

Cognitive Engineering

01 Law of Conservation of Awareness
Awareness cannot be created or destroyed, only changed from one form to another. The two forms of awareness are conscious and unconscious. Conscious awareness has no wavelength, but unconscious awareness does.

02 Unconscious memory brainwaves
Awareness generates brainwaves from active unconscious memories the same way light projected through film produces movies.

03 Cellular level brainwaves
Brainwaves are composed of millions of tiny, cellular level waves which travel through neurons the same way electric signals flow through transistors in computers.

04 Electrons flow between neurons
Neurons convey signals to one another by transmitting electrons across a gap called the synapse.

05 Genetic modification
Neurons can be genetically modified to raise their resistance to incoming electrons.

06 2nd Law of Neurodynamics
Raising a neuron's electrical resistance diminishes electron flow and attenuates unconscious brainwave activity. A person's unconscious brainwave field is joined in an electrical circuit with their body's neurology.

07 Higher states
Reduced unconscious brainwave activity restores greater free conscious awareness to the individual.

08 Supporting experiments
These principles are conclusively proven by a large body of scientific evidence.

Cognitive Engineering

● **R&D Leveraging**

The worldwide investment in life sciences genetic engineering programs is generating a universal set of CRISPR tools which are transferrable to innovative applications for human cognitive enhancement. Cognitive neurogenetics redirects the CRISPR tools being developed for life sciences applications to enhancing cognitive ability in healthy individuals.

Applications: In the same way transistors were applied across many different fields, genome editing will be used in many diverse human applications. These applications fall into two major categories:

Correct: Applications in medicine to correct genomic defects
Enhance: Applications in human enhancement to increase abilities

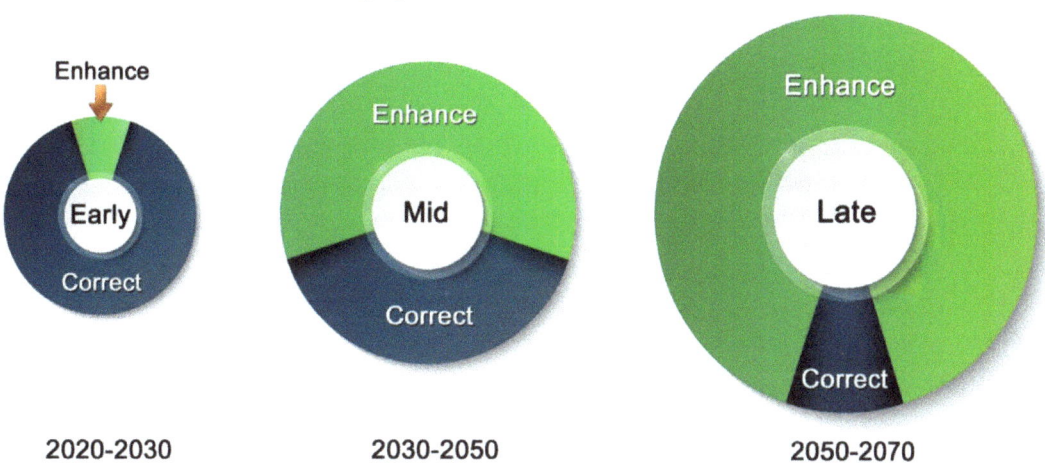

As shown in the diagram, the early applications will be primarily in medicine to correct genetic defects. The same technology which is used to for corrective applications can be redirected to improve human abilities; it is merely a matter of which genes are edited. The genome editing tools themselves are indifferent as to which gene they are applied to. Over time, human enhancement applications become dominant. In Computer Era parlance, medicine resembles the mainframe of genetic engineering, while enhancement is the internet.

With $22 billion already invested in genetic engineering start-ups, the low hanging fruit in corrective applications is gone. New projects will increasingly have to tackle multiple genes or genes which are difficult to edit. Enhancement is the new frontier.

Cognitive Engineering

● Cognitive Economics

Artificial intelligence and robotics will displace millions of workers across a wide range of industries in the near future. As these new technologies take hold, they will also create millions of high-paying, cognitively-demanding jobs. In today's interconnected world, these new jobs will inevitably emerge in countries where the mental capacity of the workforce matches their demands. In short, we are entering an era of global cognitive economics.

● Cognitive Geopolitics

The new global economy is an innovation race which China is winning. The worldwide center for microelectronics has already migrated from San Francisco to Shenzhen, and software is next. Silicon Valley is microdosing to stay ahead.

To thrive in this century of global competition, American business needs more innovation, more creative solutions, and more exceptional thinking.

World-renowned Oxford scholar Nayef Al-Rodhan believes that harnessing cognitive enhancement technologies can help nations engineer more productive, focused and competent workforces; thus raising the overall output of their economies and projecting greater global power. (1)

China is moving ahead aggressively in human genetic enhancement and is now the world leader. Darryl Macer of the Eubios Ethics Institute predicts that China will be at the forefront of human genetic enhancement. Since Western countries have more conservative attitudes about synthetic biology, he argues that China is set to lead the world in genetic enhancement. (2) By upgrading the cognitive abilities of its population, China could become even more competitive on the global stage, while American businesses close their doors in vast numbers.

Global cognitive economics is a high-stakes game which demands bold thinking and dynamic action. China has already genetically engineered 86 people, and this is only the beginning. (3) To protect democratic free economies from an onslaught of cognitively-augmented Chinese workers, Western leaders must plan ahead and start taking action now.

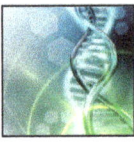
Cognitive enhancement represents a much more than a lucrative investment opportunity. It is an economic imperative to safeguard free world economies and uphold the values and ideals we cherish.

Cognitive Engineering

● The Genetic Age

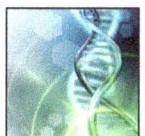

The previously-unfathomable mechanisms of gene interaction have been penetrated by a revolutionary technique called CRISPR. This development is a monumental landmark in the history of biological research. Its impact on human civilization is being compared to the steam engine, which gave rise to the Industrial Revolution, and the transistor, which spawned the Information Age.

Genetic engineering can be used to enhance cognitive faculties such as mental acuity, awareness and intelligence, and physical attributes such as appearance, strength, and agility. Investment in gene therapy is booming. According to The Alliance for Regenerative Medicine, public and private genetic engineering companies raised $22 billion globally in just three years.

Humanity is standing at the threshold of a Genetic Age. One way to understand the progress which awaits us is to compare this new age to its predecessor, the Information Age.

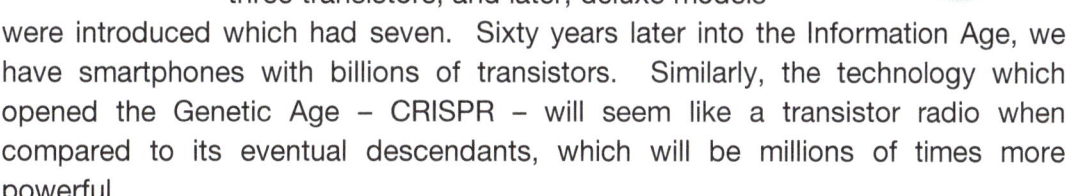

The first portable entertainment device to appear during the Information Age was the transistor radio. The early models had three transistors, and later, deluxe models

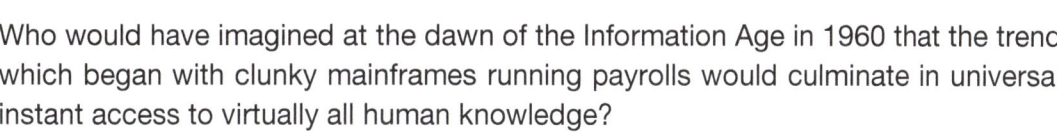

were introduced which had seven. Sixty years later into the Information Age, we have smartphones with billions of transistors. Similarly, the technology which opened the Genetic Age – CRISPR – will seem like a transistor radio when compared to its eventual descendants, which will be millions of times more powerful.

Who would have imagined at the dawn of the Information Age in 1960 that the trend which began with clunky mainframes running payrolls would culminate in universal instant access to virtually all human knowledge?

Cognitive Engineering

Humanity is facing a similar dimension of opportunity as it stands at the cusp of the Genetic Age. Genetic engineering can lift millions of people out of darkness into higher states of being, leading to a democratization of awareness much the same way as the internet has created a democratization of knowledge. The human race can use this newfound power to expand its cognitive capital to provide the wisdom necessary to safely manage the exponential growth of technology.

● Wisdom Gap

The 21st Century will see a constant onslaught of change and an ever-increasing flood of new technologies, some of which may be more powerful than anything we can imagine today. The speed of scientific advances is multiplying exponentially, and the potential for abuse in robotics, nuclear energy, artificial intelligence, and other powerful technologies is beyond our capability to measure.

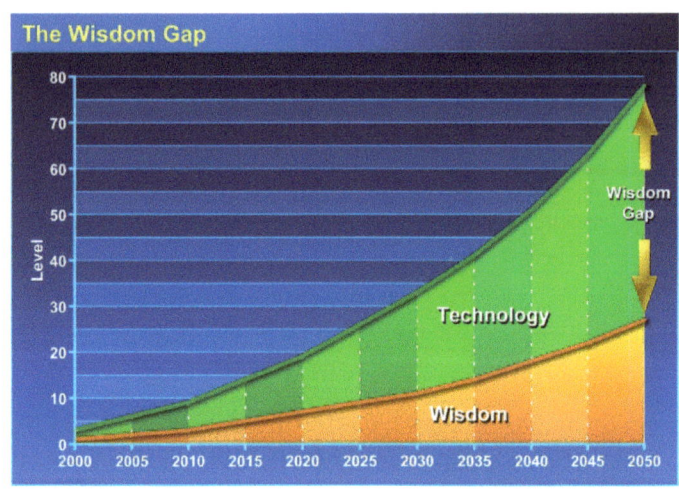

These new technologies are primarily under the control of industrialized nations which have not demonstrated the ability to manage scientific innovation wisely. Our current leaders and policy makers have a narrow perspective which focuses on immediate payoffs without considering long range impacts – creating serious, widespread risks which may jeopardize humanity's future.

Microprocessor power is the driving force behind advances in almost every field of science and engineering, and it is growing exponentially according to Moore's Law.[14] Since there is no counterpart to Moore's Law in metaphysics or the wisdom sciences, the disparity between progress in the physical and human sciences is growing wider, creating a "wisdom gap" as illustrated in the graph. As the rapid pace of advances in the physical sciences outstrips humanity's capacity to manage technology wisely, the wisdom gap carries increasing risks of misusing ever-more powerful technologies.

---

[14] Developed by Intel co-founder Gordon Moore, this law states that computer power doubles every 2 years.

In the last century, the risks of misapplying scientific breakthroughs in atomic energy escalated, culminating in a Cold War which produced enough thermonuclear warheads to annihilate humankind many times over. Today, the misuse of energy generation systems is creating climate change problems which could potentially endanger life and cause damages in the trillions of dollars. Other examples are shown in the chart.

Misapplying Technology		
Technology	Misapplication	Result
Energy	Climate change	Imperils biosphere
Physics	Nuclear warheads and reactor meltdowns	Endangers human life and society
Derivatives	Financial crisis	Global recession
Entertainment	Violence in movies, books and video games	Violent behavior in young people
Pharmaceutical	Harmful medications	Toxic effects
Software	Viruses	Needless cost
E-Trading	Casino capitalism	Market volatility
Internet	Hacking	Security threats

Given the potential consequences of misapplying revolutionary new technologies, humanity would be well advised to mitigate its risks by taking deliberate action to close the wisdom gap. To ensure humanity's future, the consciousness of decision makers in developed countries must be raised to higher levels which yield broader perspectives and wiser decisions.

The technologies disclosed in this book can anchor a global wisdom initiative to raise humanity's cognitive capital. Such an initiative can close the wisdom gap between physical and human sciences, and enable neuroscience to keep up with the accelerating pace of advances in technology and physical science. This will reduce the risks of misapplying powerful technologies and protect humanity's future.

● Safeguarding Humanity

Futurists agree the 21st Century will witness a continual barrage of scientific breakthroughs and technical innovations. The blistering pace of advance is outstripping humanity's ability to manage technology wisely, placing potent new technologies in the hands of business leaders and policy makers who are fixated on short-term results without considering long-range impacts. This combination of scientific prowess and narrow-minded shortsightedness is creating serious risks which may imperil humanity's future.

Through a global initiative to expand conscious awareness levels around the world, humanity can enrich its collective wisdom to mitigate these risks and safely manage the exponential growth of technology in the 21st Century and beyond.

Elevated awareness opens people up to accepting the principles of wisdom. As countries across the globe adopt these principles, humankind's collective cognitive capital will rise to counterbalance the risks of runaway technology change.

With the expanded perspective higher consciousness provides, policy makers at all levels will start considering the long-range effects of their decisions. Policies forged out of greater awareness will safeguard humanity's future.

- Changing History

Global evolution to higher states of consciousness can open a golden age of wisdom for humanity, toppling materialistic paradigms in the West and reinvigorating spiritual traditions in the East. These changes will create an explosive demand for spiritual guidance and education, accelerating human conscious evolution.

- Ensuring the Future

Raising world awareness of higher states of being will tend to galvanize public support of long-range initiatives, such as sustainable energy and environmental protection. Policy makers at every level will start to give greater consideration to the impact of their decisions on future generations.

This shift into higher consciousness will work to defuse irresponsible short-range public and corporate policies which emphasize near-term results at the expense of long-range sustainability. As thousands of key decision makers around the world start to see themselves as inheritors of the future they create, they will forge more enlightened policies which will protect humanity's future.

- Shifting Behavior

Industrialized societies have an urgent need to change the underlying values that are feeding a culture of frenetic consumption which is clearly unsustainable. The global expansion of higher awareness will shift household spending patterns and capital flows as consumer focus begins to move from acquiring temporary material possessions towards accumulating lasting spiritual wealth.

In addition, the worldwide growth of spiritual awareness can liberate humanity from the fear of death, altering end-of-life care decisions and priorities. A global shift to more empowered attitudes towards death and future lives will raise worldwide demand for greater legal freedom in end-of-life choices, reducing needless suffering and expense.

● Youth

The scientific engineering of consciousness can re-engage younger generations into the pursuit of wisdom. Today's young people are tomorrow's leaders. The technology's exciting results can prevent an erosion of humanity's precious spiritual capital by counterbalancing the mesmerizing effect of luxury and affluence on young adults.

● Cognitive Capital

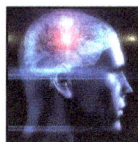
We are witnessing the emergence of a new kind of wealth – genetic capital. This new form of capital will create value in a myriad of applications across agriculture, energy, health and industry, but perhaps its greatest potential for value creation is in cognitive capital.

The world today is shifting into a knowledge economy where cognitive capital is the prime currency. Humankind's cognitive capital today is presently tied up primarily in unconsciousness. In the 21st Century, the ability to liberate cognitive capital – conscious liquidity that can be applied to creating new solutions – will determine the fate of individuals, communities, corporations and nations.

● Leadership

Many challenges lie ahead, but there is considerable cause for optimism about the potential for consciousness engineering technologies to drive widespread changes in human knowledge and behavior which can generate substantial benefits across the globe.

Visionary leadership will be required to achieve the full promise of these scientific and technical innovations, but the potential benefits are enormous. It will be up to progressive spiritual, social, civic and business leaders to maximize the opportunities of these transformative technologies while addressing their challenges.

Forward-thinking leaders can be on the winning side of change by becoming early adopters and innovators who turn potential into opportunity. Well-informed policy makers who understand how humanity can benefit from these new technologies can harness their vast potential to drive the global expansion of wisdom.

Conclusions

Genetic engineering programs for enriching human awareness will present many challenges, but the obstacles are considerably less daunting than the ones NASA faced in 1962 at the start of the Apollo program. With a comparatively miniscule investment, these R&D programs can produce results in mass consciousness comparable in stature to the paradigm shift which followed humanity's first footsteps on the Moon.

Civilization has changed over advancements in technology which are more modest than the ones presented in these pages. The internet gave us a democratization of knowledge or intellectual capital. Cognitive engineering can deliver a democratization of awareness or cognitive capital.

In this century alone, billions of people have spent trillions of hours living life on automatic pilot, submerged in thought. Widespread application of cognitive engineering can restore self-awareness to every one of these lost moments, reclaiming sovereignty, freedom, empowerment and connection to Source. Over the long arc of history, imagine the impact this can have on human civilization! The achievement of permanent higher states of being can open a golden age of wisdom for humanity, and create a new generation of superconscious individuals who form the vanguard of enlightened leadership in the arts and sciences.

9

Epilogue

On July 13, 1969, the Apollo 11 spacecraft was perched on its launching pad at Cape Kennedy, ready for flight.

Three days later, the Apollo 11 astronauts blasted off for their historic journey to the Moon.

When Neil Armstrong and Buzz Aldrin landed on the Moon on July 20, 1969, they opened a new chapter in human history, and Mankind became a space faring species for the first time.

Future humans will look back and divide their history into two eras – before we ventured into space, and after.

On July 13, 1969 – literally at the dawn of Mankind's entrance into planetary exploration – a nineteen year old boy had a six hour out of body experience during meditation.

He reached hundreds of times his normal level of awareness, and actually left the physical universe.

Although a number of people have managed to get outside the physical universe, few have ever returned.

Read the following remarkable account of one of the most astonishing out of body experiences ever recorded, and gain priceless insights into your own spiritual nature.

Introduction

This chapter contains a first-person account of the extraordinary circumstances surrounding the discovery of the axioms, laws and formulas in this book.

Millions of people today have awakened to the realization that they are conscious beings, rather than their bodies. Many people treat this knowledge as if they have found the ultimate secret. Nothing could be further from the truth. Discovering our spiritual identity merely places us at the trailhead of a long path to higher and grander states of being.

The first step to achieving these higher states of consciousness is understanding the true dimensions of our nature as conscious beings. Consciousness is like a one hundred story building. In most human beings, the lights are on in the bottom five floors, and the rest of the floors are dark. Imagine the building completely lit up from top to bottom.

Unless someone has already experienced it in their lifetime, it is not always easy for people to understand what it feels like to be a conscious being who is fully outside its body. Explaining this is like trying to get someone who can only see black and white to understand what the world looks like in color. The person does not have the mental model or conceptual framework to process the new ideas.

At higher states of being, awareness, light and pleasure become indistinguishable from one another. As one ascends into these states, the bliss becomes indescribable beyond measure.

Here is a way to try to get a sense of the joy of experiencing these states of being. Suppose your favorite uncle is a billionaire, and one day, you meet with him, and you are somehow able to catapult him into the full power of the highest state for ten seconds. When he comes back, his life will be changed forever.

Supposed you then say to him, "My dear uncle, I am permitted to grant you up to two more minutes of this experience for the price of one million dollars per second. That's one hundred and twenty million dollars for the full two minutes. How much time would you like to have?" And no matter who he is (or what his values were before) if he went there, he would say without a moment's hesitation, "Sold. I'll take all of it right now." That is the power of the highest states.

Cognitive Engineering

Class Reunion

I returned to Long Island in the summer of 1988 for my 20th high school class reunion. Before the reunion, I drove to Bay Shore to visit the house where I had a 6 hour out of body experience 19 years earlier at age 19 in 1969.

On July 13, 1969, I had occupied a second story guest room which faced the back yard. I had revisited the house several times since then, but never ventured into the yard. This time, I wanted to go around into the back yard so I could see the guest room window.

The family who lived in the house 19 years earlier had moved away. I arrived around noon, hoping to find no one at home. When I drove up to the front of the house, there were no cars in the driveway, so I got out and went around the side to the back yard. The lot was much larger than one would expect for a modest house like this. It could easily have held a garden party with a hundred people.

I spent ten minutes walking around the back yard, looking at the guest room window from different angles and taking pictures. As usual whenever I had returned, I felt a deep appreciation of the sacred place I was in. After a while, I decided I wanted to touch the house.

(I should mention that in my life, I have known "sensitive" people who are psychic and can sometimes see "energy" or auras. However, I was never one of them.)

I slowly approached the house with my outstretched hand. As my fingers came within a few inches of the surface, something astonishing happened. What looked like a large bolt of orange electricity sparked out from the house into my hand, accompanied by a loud crackling noise that sounded like a thunderbolt. The bolt was about three inches thick, and it glowed orange like the heating elements inside a toaster. It was so powerful it nearly knocked me over. In that moment, I knew I was witnessing a faint psychic residue of my experience there 19 years earlier.

July 13, 1969

Cognitive Engineering

After I went off to college in the fall of 1968, my sister, Maggie, was elected editor of our high school's literary magazine. She attended a convention for literary magazine editors, where she made friends with a girl named Rita from Bay Shore. (Bay Shore is about 20 miles southeast of Syosset on Long Island, where our family lived.)

When I came home from college on spring break in March, 1969, my sister, mother and I visited my sister's new friend, Rita, and her brother, Mark, who was also home from college. During this visit, we happened to see a rare, breathtaking, full-color aurora borealis. In retrospect, it was evidently a sign.

Four months later, in the early summer of 1969, I was home from college working two jobs. The first job was making vacuum cleaner parts in a plastics injection molding factory from midnight to 8:00 AM, and the second one was stocking shelves in a plastic bag warehouse from 9:00 AM to 1:00 PM. Our family home had a finished basement and I slept there during the day.

Early in July, Rita had invited my sister and me to a party she and her brother were planning for Sunday, July 13. This date happened to coincide with a camping trip which my parents and sister were taking in Vermont, so my sister could not attend. Since I was working, I did not go on the camping trip, and I was free to attend.

On Sunday, July 13, 1969, I got up at 5:00 PM and drove my mother's black Ford out the Long Island parkways from Syosset to Bay Shore. I arrived in Bay Shore at 6:00 PM and greeted Mark and his sister, Rita. They had the house to themselves and a small party was going on in the living room. Mark and I fixed snacks and soft drinks in the kitchen and brought them out to his guests in the living room. After a while, we went up to his room and he showed me several books by Edgar Cayce. Dr. Cayce was a phenomenally gifted psychic who came to international prominence in the 20th Century and, like many people, Mark was captivated by Edgar Cayce's amazing abilities.

Eventually, we went back downstairs to the living room and joined Rita and her friends, who were talking and listening to records. This went on until around 11:30 PM, when they left to go to another party. That left Mark and me... and some LSD.

It was a balmy summer night, so we decided to go outside. We went on the front lawn and played frisbee under the light of a mercury vapor street lamp. It was at this point that I first left my body. I was animating my body playing frisbee while being partially external to it. I felt very free and exuberant. We threw the frisbee around for fifteen minutes, and then decided to take a walk to a nearby harbor to see the boats.

The harbor was an inlet of water about a mile long and 500 feet wide which opened up into the Atlantic. There were hundreds of boats moored on both sides of the inlet. Access roads and houses ran up and down each side of the harbor.

At the landward end of the inlet was a small park with two pavilions separated by a grassy area with a flagpole. Each pavilion had a cement floor encircled by six white columns supporting a green tile roof. A spotlight shone down from each pavilion's roof onto the cement floor beneath.

The combination of tile roof and cement floor yielded acoustics which created a strange and subtle echo to sounds. When Mark and I talked in the pavilion, it sounded like we were speaking into a microphone, and when we stomped our heels on the floor, it made a staccato echoing noise like the reverb unit of an amplifier.

At the head of the inlet, the land around both sides of the water sloped up into low, grassy hills dotted with trees. We sat down on the east side and surveyed the harbor. The full moon was reflecting brightly off perfectly still black water, silhouetting the colors of the many boats moored there. I felt absolutely serene.
I noticed I was in spirit-to-spirit communication with the tall green trees on the other side of the inlet. I saw and felt the life energy of the trees like a luminous green cloud glowing softly against the night sky.

I rose and looked out to the ocean. I was standing on a moonlit hill; sea ahead of me, and to my right, a harbor with boats sleeping. The harbor scene was paper thin, like a still-life portrait in front of me. The sparkling ocean flowed over the faint horizon out into the glittering black sky. A gentle breeze warmed the air to my touch, and night's blanket enfolded sea, stars, harbor and me. I was completely and beautifully detached from everything I saw, and I felt the life in a thousand stars.

I had very few thoughts, but I remember the ones I had. They ran something like this:

Night is the alternate world of starry skies by still waters where you can feel yourself exist and face the question of "What Is Reality?" as it stares back at you.

The differences between Earth and other planets are blatant in bright sunlight, and subtle in cool moonlight. In the night time, things blend into one another. Space is shortened, and time evaporates.

Night cities from the air could be cities anywhere. Differences between objects, like expensive and inexpensive cars, are much less pronounced at night. They become cars in the night.

Things are as if underwater, unified by a calm blanket of balmy breeze in the warm darkness.

It is easy to find places that are not crowded or noisy at night. There is room to expand in, and you can own all you survey.

At night, there is unity with other life in the galaxy. It is easier to lose time and location at night. You could be anywhere, any time. You could even be on another planet. So you discover you are nowhere, and now you have truly arrived. You have arrived at Nowhere, your final destination, and you have reached the point where philosophy actually begins.

After soaking in the harbor scene for a while, Mark and I decided to walk back to the house. Mark said he felt like jogging, so I increased my pace to keep up with him. As I moved faster, I felt myself going back into my body.

A few minutes later, we arrived back at the house. As we were going in the front door, Mark's dog, Tippy, escaped, and he went out into the neighborhood looking for her.

I went up to a guest bedroom on the second floor and sat down against a wall. The room had one lamp which had a red bulb and a colored scarf draped over the lampshade. It bathed the room in a warm hue which made it feel like a very safe place. I was alone in a warm and comfortable place surrounded by silence. I felt calm and serene, and went into a deep meditation.

My body was sitting cross-legged against the west wall of the room. Its torso was arched over, bending forward, completely relaxed. Its head was facing down and its eyes were closed.

Astonishing realizations began erupting through my consciousness in continuous succession, one after another. I had the thought that if I wrote the ideas down, I would have a trail back to the higher states of being that I could follow any time. I decided, however, that I was never coming back to Earth, and so there was no need to write anything down.

I climbed, and climbed. For the rest of the night, I would be in a state of being which can only be truly known by those who experience it. I filled the entire room and it became brilliant. I saw and had full perception not even distantly related to the human body propped against the wall.

Sages throughout the ages have been at a loss to describe exalted states of being in words. But a picture is worth a thousand words, so why don't we try that?

This illustration shows what a really terrific, top-of-the-line, out-of-body experience can look like. The spiritual being is connected to its body by a silver cord. The being is surrounded on all sides by space (instead of being inside of a solid human body). You can imagine how wonderful this freedom would feel.

A being in this state can have one hundred times its normal level of awareness. To gauge this, imagine how it would feel to have ten times your current level of awareness-of-awareness. Then — multiply that by ten. Imagine yourself in this picture. You are the radiant sphere of awareness floating in the space of the room.

I was a radiant sphere of awareness floating in the center of the room. I was around four feet above the body and three feet in front of it. A thin silver cord tethered me to the back of the body's head. Other than this cord, there was no longer any connection between me and the human body slumped against the wall –

or the race of people it belonged to, or the planet they lived on. That life was gone; vanished, as if it never existed. It was not history, nor was it a memory. It just disappeared.

Picture a small room with a ten thousand watt light in its center, piercing into the space in the room and illuminating it brilliantly. Imagine the light is not light energy streaming into space; rather, it is pure awareness; pure you. You are awash in light, and your body's eyes are closed. Awareness is light. Spirit is awareness. Consciousness is the light of the universe.

At the highest states, everything goes. You can't move, think, speak, read or write. You can do any of these things, but you know you that to do them, you have to trade off free awareness to perform the activity, and the only thing you want is more free awareness, not less.

Higher and higher I went. I saw Earth Reality as a ludicrous illusion which dissolved and fell away beneath me. I experienced magnificent, joyous, radiant awareness. I reached out into the galaxy and saw it teeming with life. The luminescence of life throughout the universe was sparkling.

Other than the amazing insights, I had no thoughts. Thoughts disappear long before the most rudimentary external states begin. My only thought was   I   A M.

As I rose up through successive planes of awareness, I encountered two gigantic beings of light. Although my parents were agnostic and I had no formal religious education, I recognized the first being instantly as Christ.[15] The second being had an Eastern presence which I perceived as Buddha. The size of these beings was absolutely enormous. I was a sphere of awareness the diameter of a beach umbrella, and they were the size of Ferris wheels (at least, the aspects I could perceive).

As I continued to climb, I found myself looking at a helix-shaped object. It was three feet tall with a glassy surface and thousands of small facets radiating every color in existence. The body's eyes were closed and I was completely outside it. I saw the object myself directly. In retrospect, I believe I was looking at what I would have to call a spiritual DNA template; permanent memory of the information frequencies which coalesce into physical DNA.

---

[15] There is a backstory which will interest theologians, but since this is a science book, I will elaborate elsewhere.

Incredible realizations erupted through my entire beingness and sent me blossoming upwards over and over. I recovered all of my awareness from the past and came fully into present time. All connection between me and my former life, body and incarnation was dissolved. I was free.

Outside the Physical Universe
When I left the physical universe, I was a bubble of awareness the diameter of a beach umbrella floating in a sea of awareness. I was pure awareness inside the bubble, and pure awareness surrounded me outside the bubble. The bubble membrane was my identity as an individual being.

I was in a field of awareness extending infinitely in all directions, and it was self-aware. This was Divine Intelligence, the body of God. The physical universe was far below. It looked like a large, sparkling diamond against a backdrop of black velvet.

The Bubble
An external conscious being who is surrounded by space (instead of a body) is like an air bubble in the ocean. Space feels as dense to the conscious being as seawater does to the air bubble.

The bubble is not part of the ocean; it is an ambassador from the atmosphere above. No part of the air inside the bubble is wet. The bubble is a place where the water is not. It is a hole in the ocean. The ocean forms itself around it.

The free conscious being is not part of the field of space which surrounds it. It is an envoy from the awareness universe above. The being is a place where space is not. It is a hole in space. No part of the being is in space. Space forms itself around it. Space is like a thick blanket of air around the being; it is much denser than awareness.

There is a border or membrane between the being and the surrounding space. It is the boundary where a conscious being ends and space begins. It is impossible to detect this boundary in embodied states, but it is quite plain in very high bodyless states.

Spirit / Space Membrane
A conscious being who leaves the physical universe is like a soap bubble which emerges from the ocean and rises into the air. The soap bubble's membrane is the being's identity. If you pop the bubble, the being ceases to exist as an individual identity, and its composite awareness merges with the main body of God.

Cognitive Engineering

● Into the Void

Above the highest point of awareness attainable as a conscious being lies the ultimate destination, where one relinquishes individual identity and blends indistinguishably into "nothingness," the Void or God, and no longer has an identity or an individual existence. There was a period during my meditation where I was within reach of ceasing to exist as a being and phasing into non-existence, living only through God. Only a tiny thin thread associating me with a body stopped me from floating off into space and leaving it to die with a perfect smile on its face. Except for that most delicate of threads, it was within my reach to phase up into the Void, cease to exist as a beingness, and become indiscernible from God.

● Describing Bliss

In the beginning of this chapter, an example was given to try to communicate a measure of the joy of experiencing the higher states of being. We imagined what would happen if you were able to snap a billionaire uncle of yours into the full power of the highest state for ten seconds, and then offer him two more minutes more for the price of one million dollars per second.

We posited that your favorite uncle would say "Sold. I'll take all of it right now." It is actually better than that. The distance between the highest state of being and normal human awareness is so vast that if you only took your uncle halfway there, the result would be the same. You would not have to discount it one iota.

At higher states of being, light, awareness and pleasure become indistinguishable from one another. The experience I had was not like a lifetime of pleasure in one night. It was like a lifetime of pleasure in one moment, and the moment went on for hours.

There is no word in the English language which describes experiences like this. They are so rare that they have not found their way into the language yet. The English word which comes closest to describing how it feels to be in higher external states of being is orgasm. Human beings understand orgasms as eruptions of pleasurable awareness they experience while living inside their physical bodies. Although perfect in its own way, the human orgasm is just a pale shadow of the pleasures that await us in higher states of being.

Physical orgasms involve pleasure centers in the brain. I was not in the brain. I was in the center of the room. I had left the brain long ago.

All bliss comes from the conscious being itself. Erase the body from the orgasm equation and just imagine the feeling of pleasure happening in you as a free conscious being surrounded by space instead of by physical matter. This will give you an idea of an orgasm of beingness. Then multiply the orgasm by one hundred and have it go on for hours. That will give you an idea of my experience.

Actually, this is not the complete truth. I hesitate to mention the truth, because it is too much. I could never have conceptualized the full stature of these states right up until the time I experienced them, so I don't know how I can expect others to do so. But for the record, I will say the number was not a hundred. It was more like a thousand.

Next Day

From a totally expanded external state, the idea of being inside a body, surrounded by dense, solid mass, instead of being free, surrounded by space, was unimaginable. It was unthinkable. As far as I was concerned, I was not coming back. I did not expect to come back. I was free.

Magnificent states of being can be very stable when they are occupied by bodyless conscious beings on the angelic planes. Making these states persist for incarnated beings is much more challenging, and I certainly did not have the requisite expertise as a teenager.

My attachments to my young life on Earth were deeper than I realized, and as dawn approached, my awareness slowly began to fade. Towards morning, Mark came into my room and sat down for a few minutes by the wall opposite my body. Neither of us spoke. After he left, I laid down on the floor and looked at the ceiling as the sun rose outside.

After a while, I went out and sat in the back yard until Mark and Rita called me in for breakfast. Mark told me that after retrieving his dog, he had listened to his stereo and read Edgar Cayce books. I tried to explain to him what had happened, but I could not find the words to express it.

When it came time to leave, I could not believe I had to drive back to Syosset in the black Ford when I could be anywhere at will without the body. But I did drive back, because that morning, I had an appointment.

After having this experience, someone more spiritually mature than me would have simply abandoned their life and joined an ashram. I went to a job interview. I applied for a summer job at the post office that would pay a lot more than my other two jobs combined.

For the rest of the summer, I had the ability to leave my body at will. During the next six weeks, I would come home after work, go to my bedroom, sit on my bed, meditate, get out of my body, and float into the middle of the room. I would remain in that state for hours. Most of the total number of hours I have spent outside of my body in this lifetime were during the six weeks following July 13, 1969, meditating in my bedroom.

At first, it only took me a few minutes to get out of my body, but as the weeks went by, it took longer and longer. By the end of summer, the ability had faded. During the last week in August, I finally got around to making out with my girlfriend, Tova, on the dunes at Tobay Beach. In September, I want back to college and she left to live on a kibbutz in Israel.

Afterword

The day after I had the experience in Bay Shore, I did not abandon my life to live in an ashram, or become a monk. I went to a job interview. In the decades that followed, I had a dynamic business career, happy and fulfilling relationships, and an active lifestyle. I lived.

Throughout the ages, thousands of monks have spent millions of hours yearning for an experience of Divine union. Why was a spiritual neophyte like me given this prize while it eluded legions of other more advanced souls? Why did I receive it instead of others who were more evolved and deserving?

Half a century of reflection has provided an answer. Although a number of people have managed to get outside of the physical universe, few have ever returned. If I were any wiser or more mature than I was, I would not have come back. The only way to get this kind of information back to the Earth plane is to give it to someone who is young and stupid enough to come back. And I was the right guy for that job.

The insights I received have guided my life and research for half a century. I have set forth the key axioms and formulas of cognitive physics into this book, and used them to create a new science for uplifting human consciousness to higher levels of awareness. What will be the outcome of this work? Is there any evidence it will be widely adopted?

Cognitive Engineering

Double Aurora

When I came home to Long Island from college on spring break in March, 1969, I went out to the Hamptons with my sister and mother to visit family there for a few days. On our way back home on Sunday, March 23, we passed through Bay Shore around dusk and stopped for a visit. As I mentioned, we met my sister's new friend, Rita, and her brother, Mark, who was also home from college.

As this was my first visit to Bay Shore, I wanted to walk around and see the area. So while our mothers socialized, the four students went out for a long walk around the neighborhood. To our astonishment, we saw a rare, breathtaking, full-color aurora borealis. The aurora looked like curtains of multicolored light hanging down from the heavens, stretching a hundred miles up into the sky.

The next day, the spectacular aurora was reported in 81 newspapers.

Cognitive Engineering

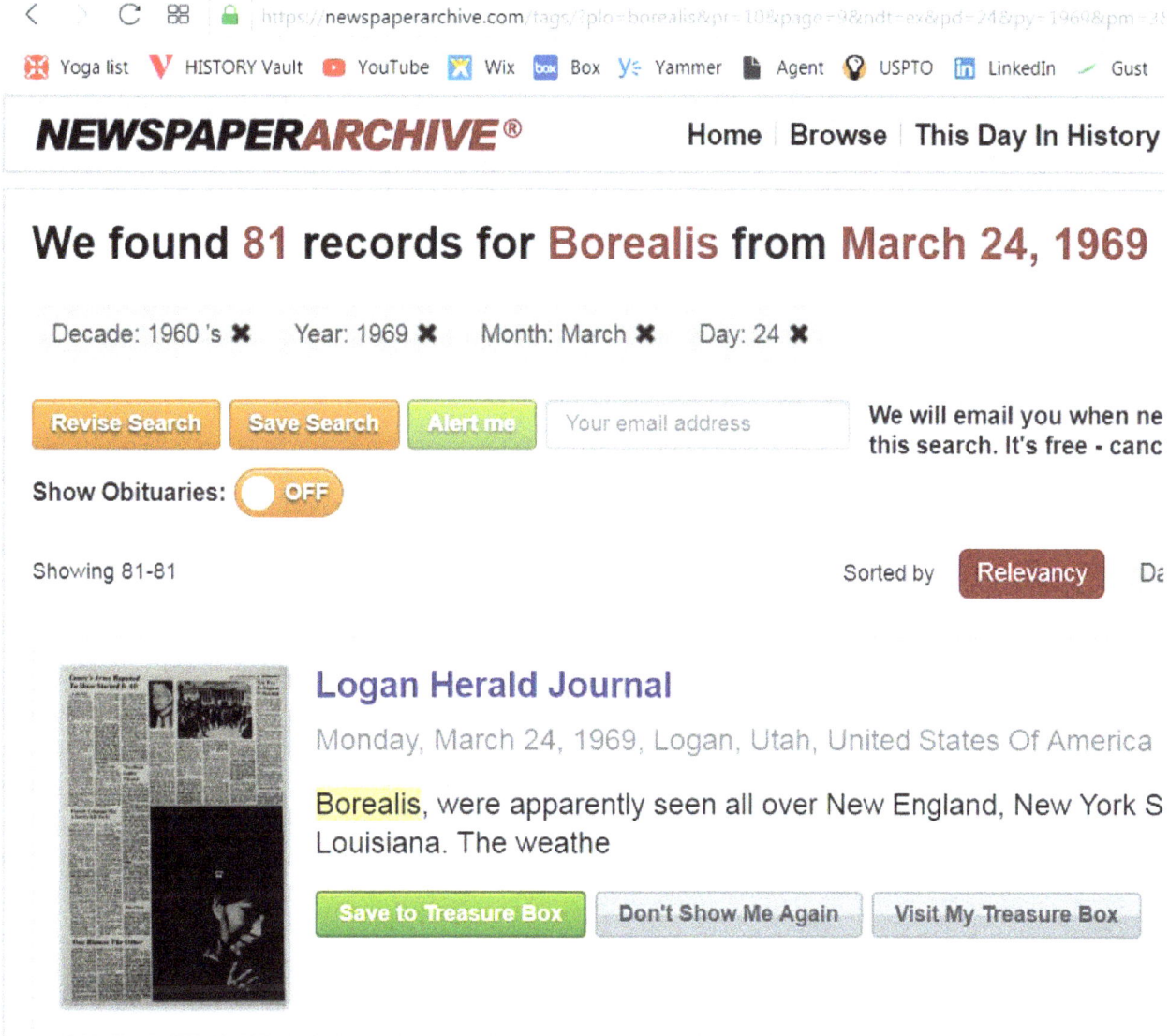

When I returned to Bay Shore four months later, I had a six hour out-of-body experience during meditation which shaped the rest of my life and formed the basis for my work.

Cognitive Engineering

● Observations

It was not likely that I would ever witness an aurora at all, as the northern lights are not normally seen in these latitudes. It was even more improbable the aurora I saw would be as magnificent as it was. (News reports called it the most spectacular display in since 1941.) However, thousands of people on Long Island saw the aurora that night, so my sighting, while unlikely, was not unique.

What *is* unusual is that I saw the aurora on my first visit to Bay Shore, and that my second visit four months later transported me into a state of Divine union. The odds of anyone suddenly entering a state like this during a lifetime at this stage in human evolution are remote to say the least. Overlaying the aurora on top of this yields a probability which is astronomical.

● Fifty Years Later

A month before the fiftieth anniversary of the aurora, I started planning how to honor the day, as recorded below in my journal on February 24, 2019. I decided I wanted to make a special entry on March 23, 2019 into *Iopen* (a new manuscript).

I also planned to spend the day in gratitude, prayer and meditation.

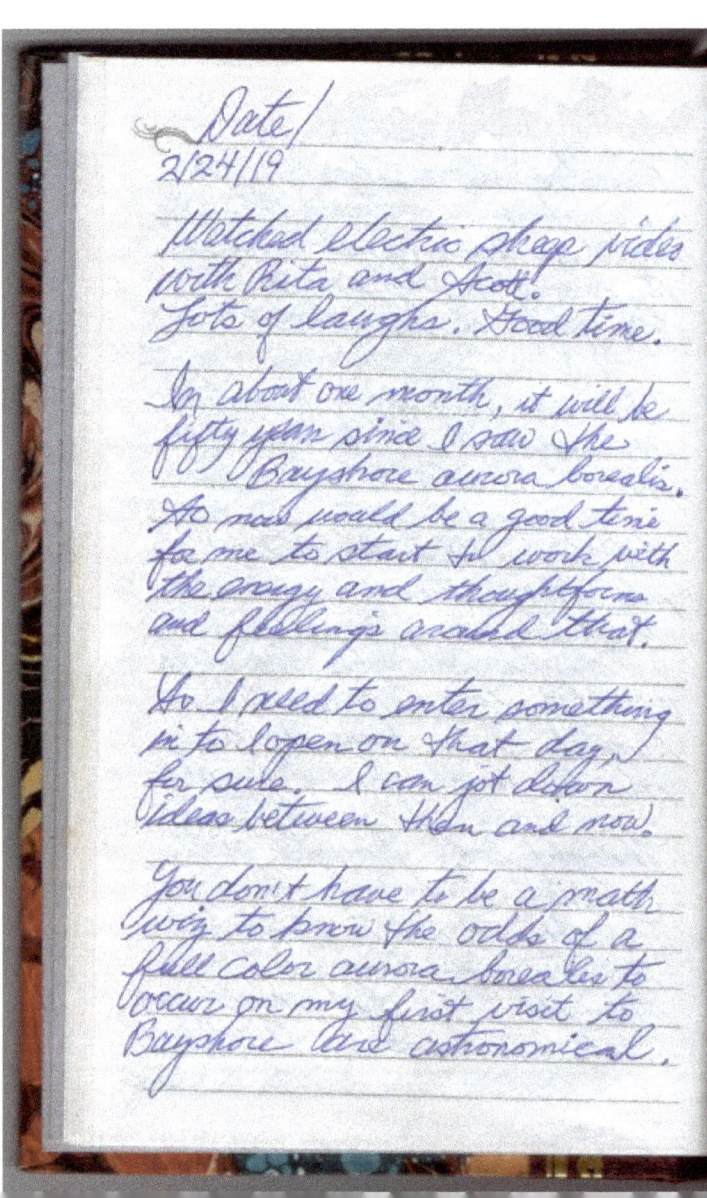

Cognitive Engineering

A month later, when I came downstairs for breakfast on the morning of Saturday, March 23, 2019 – the fiftieth anniversary of the aurora – I told my wife, Rita, I would be taking the day aside as planned.

She replied that all the major news channels were reporting the aurora was returning that night.

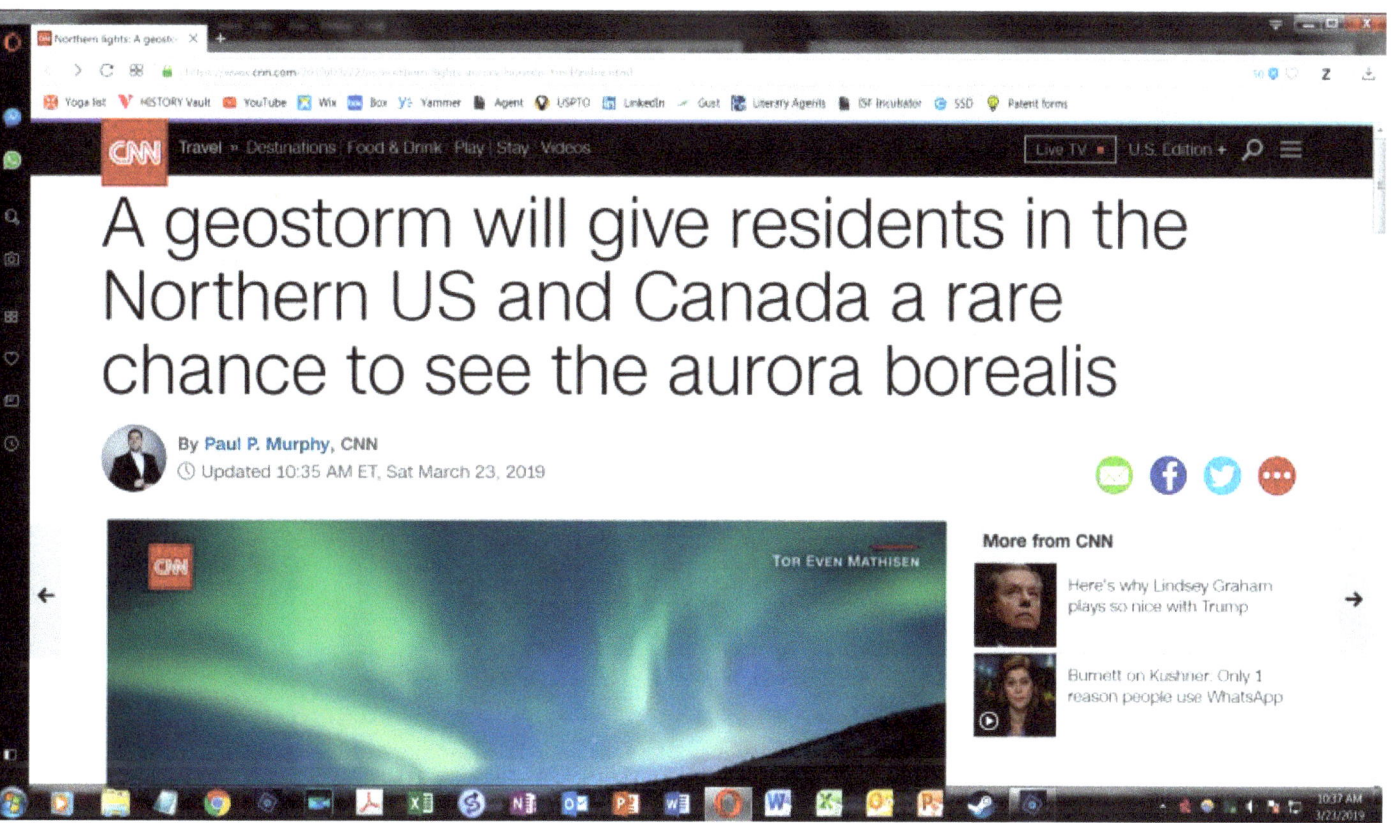

Everything stopped. I stood on the stairs for several minutes, trying to wrap my head around this. The aurora was coming back on the evening of March 23, 2019, fifty years to the day since I witnessed it in Bay Shore on March 23, 1969.

Cognitive Engineering

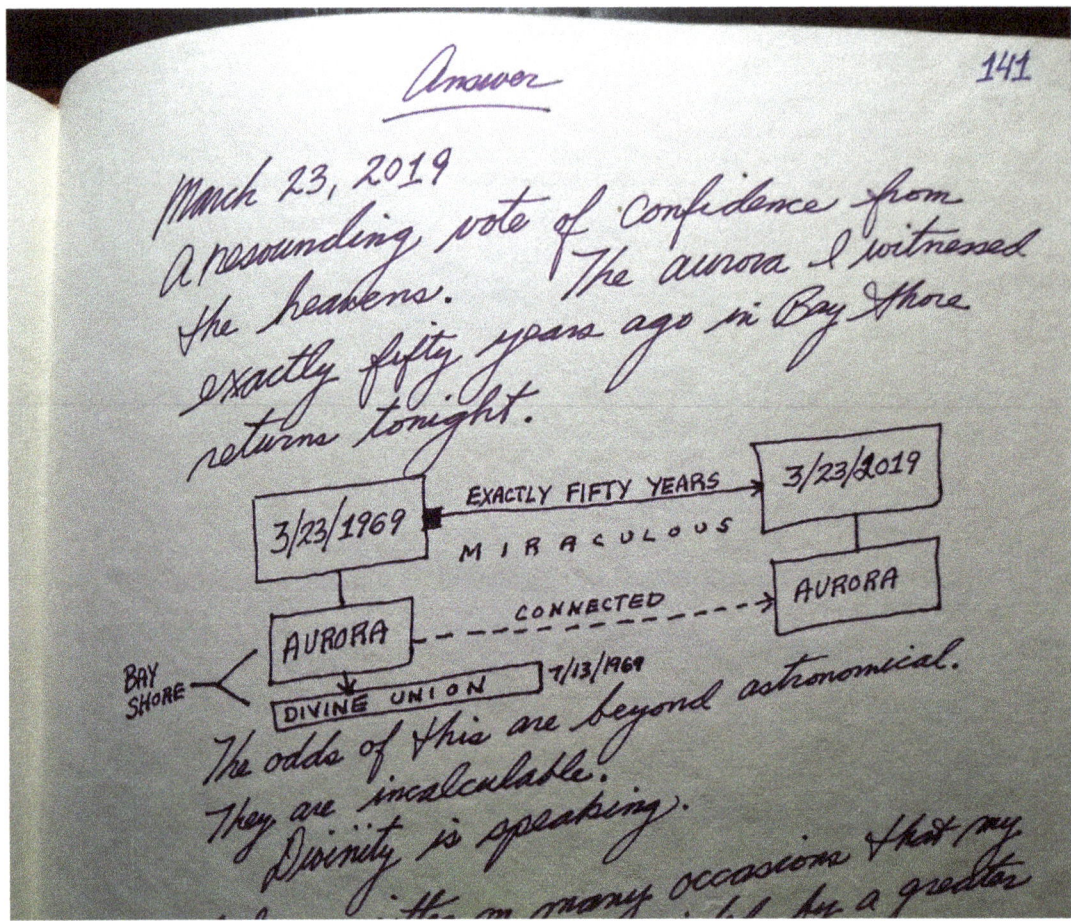

The odds of the aurora returning in exactly fifty years to the day are 1 in 18,250. If you overlay these odds on the colossal improbabilities already mentioned, the likelihood these are random coincidences becomes rather small. In fact, as I wrote in my journal that day, it becomes incalculable.

Cognitive Engineering

It could stop there, but it doesn't. The next day, I decided to relax with a good book, so I picked up my copy of *Too Like The Lightening* by Ada Palmer, which I had begun reading a few days earlier.

Delivered Mar 20, 2019

Too Like the Lightning: Book One of Terra Ignota
Palmer, Ada
Sold by: Seattlegoodwill
$2.77

[Buy it again]

I opened the book to Chapter 2, and was greeted by this:

235

Cognitive Engineering

CHAPTER THE SECOND

A Boy and His God

WE BEGIN ON THE MORNING OF MARCH THE TWENTY-THIRD IN the year twenty-four fifty-four. Carlyle Foster had risen full of strength that day, for March the twenty-third was the Feast of St. Turibius, a day on which men had honored their Creator in ages past, and still do today. He was not yet thirty, European enough in blood to be almost blond, his hair overgrown down to his shoulders, and his body gaunt as if he was too occupied with life to feed himself. He wore practical shoes and a Cousin's loose but comfortable wrap, gray-green that morning, but the only clothing item given any care was his long sensayer's scarf of age-grayed wool, which he believed had once belonged to the great Sensayers' Conclave re-

I am not making this up. As I read this page, I thought back to the morning of March 23rd — the day before — and remembered standing on the stairway thunderstruck by the news of the aurora's return.

Cognitive Engineering

So let's add this up.
1. A highly improbable series of events leads me to Divine union on my second visit to Bay Shore on July 13, 1969.
2. My epiphany is foretold by a spectacular aurora on my first visit to Bay Shore on March 23, 1969.
3. On February 24, 2019, I begin planning how to honor the fiftieth anniversary of the aurora.
4. On its fiftieth anniversary, March 23, 2019, the aurora returns.
5. The day after, I open a book to a chapter titled "A Boy and His God" which starts with "We begin on the morning of March the twenty-third…"

🟢 A Shirt From Infinity

When I drove to yoga class at the gym a few weeks later, I parked and lay down in the back seat with the doors open to rest my back as I usually do after I drive. (I have a bad back.) In the back seat, I took out my journal and wrote this:

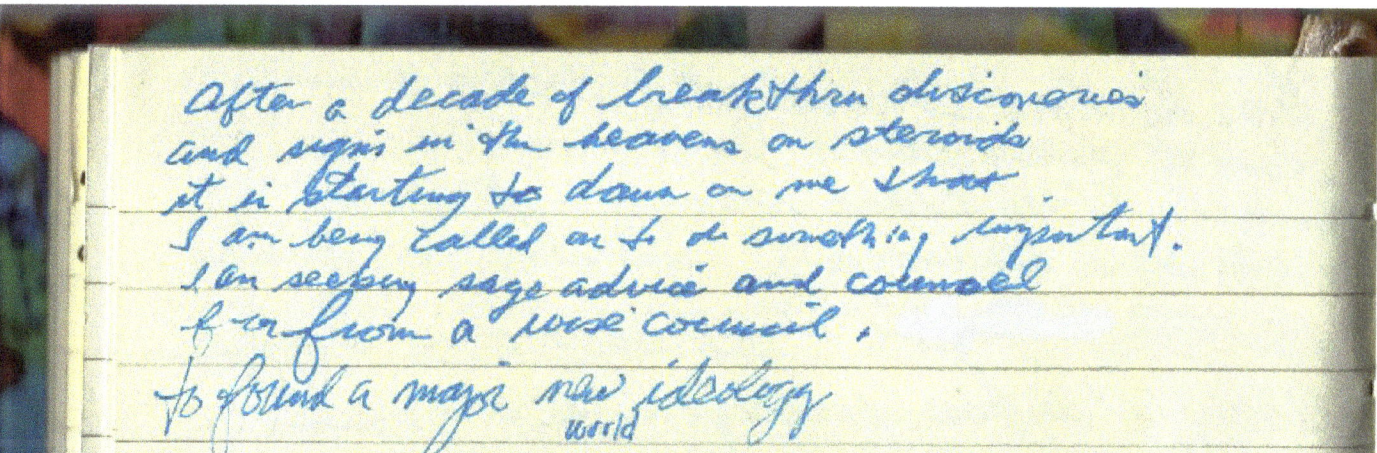

My penmanship on my back isn't great, so here's what it says:
"After a decade of breakthrough discoveries and signs in the heavens on steroids, it is starting to dawn on me that I am being called to do something important. I am seeking sage advice and counsel from a wise council, to found a major new world ideology."

After I wrote this, I thought of all the times I had laid down in the back seat with the doors open in parking lots, and recalled that only once in a decade did anyone notice. (One hot summer afternoon years ago, a young lady inquired if I was alright. I explained to her I had a bad back.)

On this day, I had parked in a different area than I usually use, and a moment later, a car pulled in beside me. A lady got out and I heard a metallic knocking sound. She was knocking on my car's fender. I rose to greet her and she handed me a package, saying "Would you like this shirt? It looks like your size." I accepted her gift graciously and she want off in the direction of the gym. She was right. The shirt was my size (S).

So a moment after writing this passage in my journal, someone materializes from out of the blue and hands me a new shirt.

It could stop there, but it doesn't. When I walked into the gym a few minutes later, I caught up with the lady who gave me the shirt and thanked her again. She seemed very happy. She explained she had bought the shirt for her son but he didn't want it. She told me her name was J.J. (As a teenager, I was called Jay.)

The yoga classroom door opened and we both entered the room to discover a substitute teacher. It was early May, spring was in bloom, and things were loosening up. Now the class usually does yoga to music, but wait a minute — that's not yoga music she's putting on. It sounds more like rock. Soft rock, but still rock. Well, no problem here. Bring it on. Substitute teacher, anything goes.

The tracks open with Procol Harum's *A Whiter Shade of Pale*, a hippie anthem if ever there was one. Then came a strange sound. What is that? Alannah Miles? I haven't heard that in decades. The last time I heard *Black Velvet* was in Phoenix. I remember hearing it in the parking lot of Papago Park in 1990, because it spoke to me then.

Papago Park

Cognitive Engineering

Black velvet and that little boy smile
Black velvet and that slow southern style
A new religion that'll bring you to your knees
Black velvet if you please…

> After a decade of breakthru discoveries and signs in the heavens on steroids it is starting to dawn on me that I am being called on to do something important. I am seeking sage advice and counsel of or from a wise council,
> to found a major new world ideology
>
> → followed by: A shirt from J.J.
> → followed by: Black Velvet

The moment I arrived back home, I asked my wife, Rita, to photograph the shirt.

I knew it was made by "IDC," whoever they were, and that was enough for me.

But my smart wife read the label and informed me who IDC was.

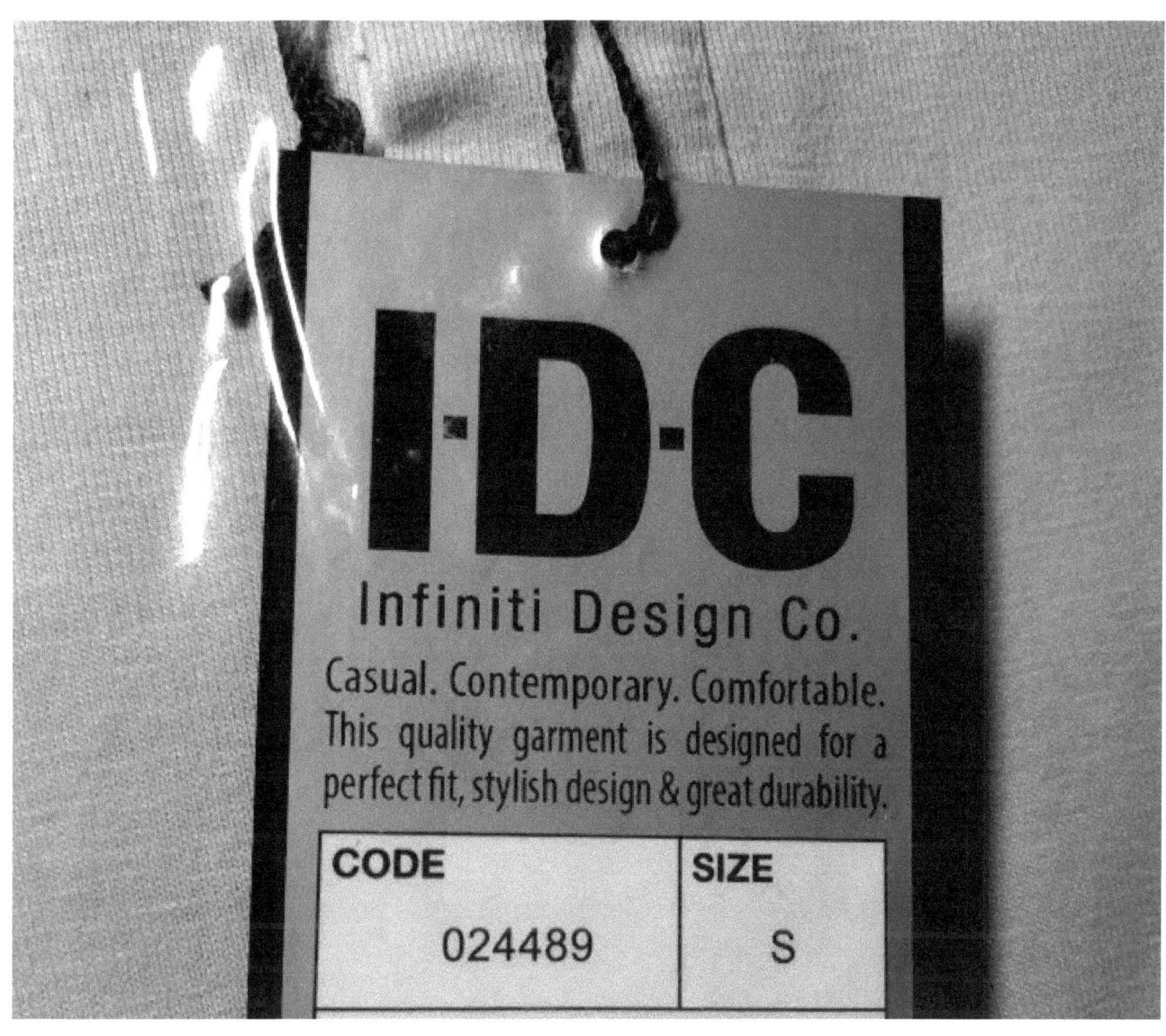

I rest my case.

COGNITIVE ENGINEERING

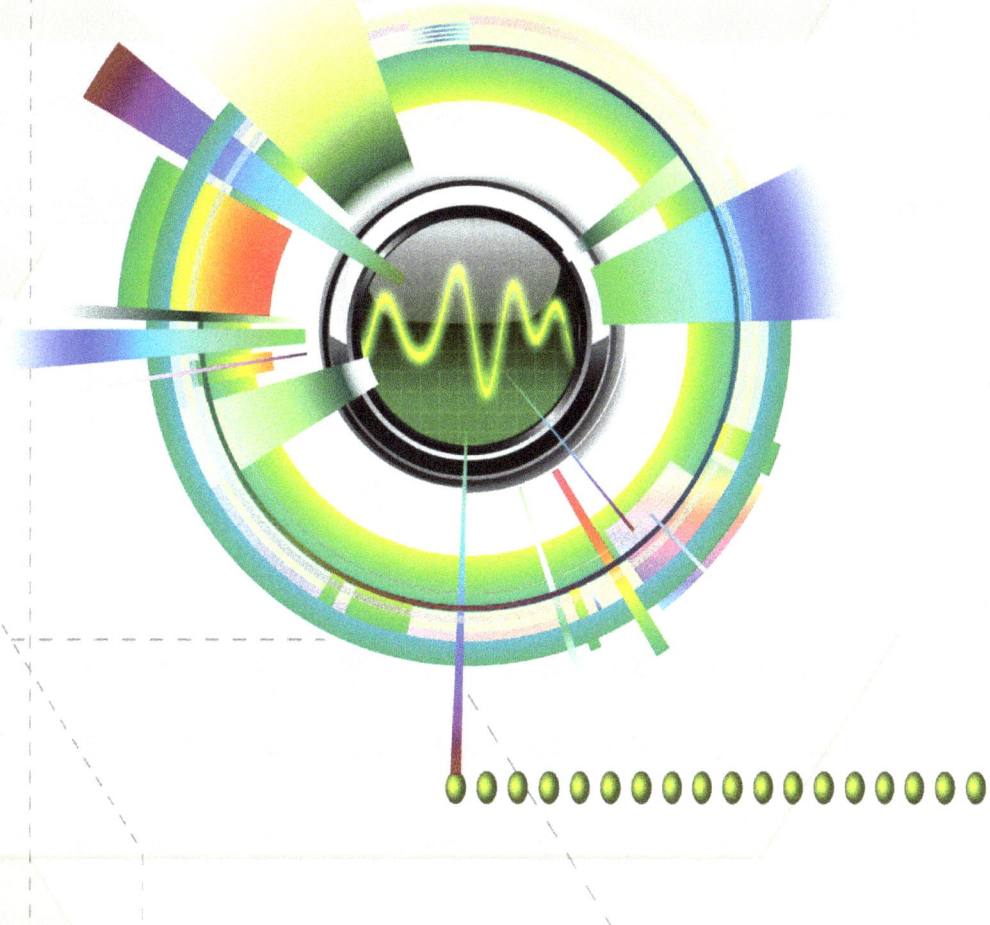

Reference

In this section...

- Summary of Axioms, Corollaries and Formulas
- Experiments
- Memory Behavior
- Neurowave Details
- Present Time and Memory
- Semantics and Nomenclature
- Biographical
- Index

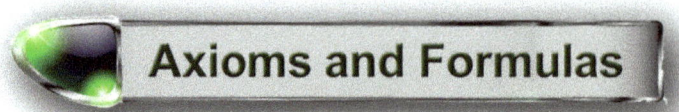

Axioms and Formulas

Group 1 – Fundamental Laws

Axiom 0 – Identity
A conscious being is a discrete unit of identity.

Axiom 1 – Awareness Dual State
Awareness has two states: free and applied.

Axiom 2 – Condensation
The living universe is a condensed form of consciousness. (Applied awareness is the first form of condensation).

Group 2 – Locality

Axiom 3 – Original State
In its original state, a conscious being has no location in space.

Axiom 4 – Locality
A conscious being occupies a location in space through incarnation.

Corollary 1 – Non-Locality
Between embodiments, conscious beings have no location in space.

Corollary 2 – Mobility
Physical distance is not a factor in the interlife.

Group 3 – Density

Axiom 5 – Law of Conservation of Awareness
Awareness cannot be created or destroyed, only changed from one form to another.

Axiom 6 – Nominal Value
Awareness elements are assigned a nominal value.

Formula 1
Law of Conservation of Awareness

$$C = (n - U)$$

where:
- C = Conscious awareness
- U = Unconsciousness
- n = the number of awareness points per being; nominally 100 terapoints

Formula 2
Beingness Volume

$$BV = \frac{4}{3}\pi r^3$$

where:
- BV = Beingness Volume
- r = Radius (= C)

Group 4 – Macro Memory

Axiom 7 – Memory States
Memory has three states: inert, active and recalled.

Axiom 8 – Memory Animation
Memory is animated by the application of awareness.

Group 5 – Neurophysics

Corollary 3 - Propagation
Awareness condenses through 6 layers to form the living universe.

Corollary 4 – Memory Power
The amount of awareness required to animate a memory is proportional to the memory's amplitude.

Laws of Perception (Radiation)

Ampere's Law
A changing electric field E generates a changing magnetic field M.
where:
Δ = amount of change
E = electric field
M = magnetic field

$$\Delta E = \Delta M$$

[Ampere's 2nd Law for Space]
A changing magnetic field M generates a changing space field S.
where:
Δ = amount of change
M = magnetic field
S = space field

$$\Delta M = \Delta S$$

[Ampere's 3rd Law for Memory]
A changing space field S generates a changing memory field Y.
where:
Δ = amount of change
S = space field
Y = memory field

$$\Delta S = \Delta Y$$

[Ampere's 4th Law for Awareness]
A changing memory field Y generates a changing awareness field A.
where:
Δ = amount of change
Y = memory field
A = awareness field

$$\Delta Y = \Delta A$$

Laws of Behavior (Induction)

[Faraday's 4th Law for Awareness]
A changing awareness A field generates a changing memory field Y.
where:

Δ = amount of change
A = awareness field
Y = memory field

$$\Delta A = \Delta Y$$

[Faraday's 3rd Law for Memory]
A changing memory field Y generates a changing space field S.
where:

Δ = amount of change
Y = memory field
S = space field

$$\Delta Y = \Delta S$$

Radin's Law
A changing space field S generates a changing magnetic field M.
where:

Δ = amount of change
S = space field
M = magnetic field

$$\Delta S = \Delta M$$

Faraday's Law
A changing magnetic field M generates a changing electric field E.
where:

Δ = amount of change
M = magnetic field
E = electric field

$$\Delta M = \Delta E$$

Formula 3
Information Density
Where:

ρ = density
i = information
sm = storage medium

$$\rho_i = \frac{1}{\rho_{sm}{}^3}$$

Group 6 – Neurodynamics

Formula 4
First Law of Neurodynamics

$$U_k = B$$

where:
U = unconsciousness value (number of applied awareness points)
B = brainwave field electric charge (newtons per coulomb)
k = a conversion factor constant (applied awareness points per newton/coulomb)

Formula 5
Second Law of Neurodynamics

$$U_k = \frac{V}{R}$$

where:
U = unconsciousness value (number of applied awareness points)
k = a conversion factor constant (applied awareness points per newton/coulomb)
V = voltage of brainwave field
R = resistance of neurology

Cognitive Engineering

(1) Meditation experience is associated with differences in default mode network activity and connectivity
Judson A. Brewer, et al
Proceedings of the National Academy of Sciences USA 2011 Dec 13; 108(50): 20254–20259.
http://www.ncbi.nlm.nih.gov/pmc/articles/PMC3250176/

(2) Mystical Experiences Occasioned by the Hallucinogen Psilocybin Lead to Increases in the Personality Domain of Openness
Katherine A. MacLean, et al
Journal of Psychopharmacology 2011 Nov; 25(11): 1453–1461.
http://www.ncbi.nlm.nih.gov/pmc/articles/PMC3537171/

(3) Neural correlates of the psychedelic state as determined by fMRI studies with psilocybin
Robin L. Carhart-Harris, et al
Proceedings of the National Academy of Sciences USA 2012 Feb 7; 109(6): 2138–2143
http://www.ncbi.nlm.nih.gov/pmc/articles/PMC3277566/

(4) Psilocybin-induced spiritual experiences and insightfulness are associated with synchronization of neuronal oscillations
Michael Kometer, et al
Psychopharmacology (Berlin). 2015 Oct;232(19):3663-76.
http://www.ncbi.nlm.nih.gov/pubmed/26231498

(5) Broadband Cortical Desynchronization Underlies the Human Psychedelic State
Suresh D. Muthukumaraswamy, et al
Journal of Neuroscience, 18 September 2013, 33(38): 15171-15183;
http://www.jneurosci.org/content/33/38/15171.full

(6) Topographic pharmaco-EEG mapping of the effects of the South American psychoactive beverage ayahuasca in healthy volunteers
Jordi Riba, et al
British Journal of Clinical Pharmacology 2002 Jun; 53(6): 613–628.
http://www.ncbi.nlm.nih.gov/pmc/articles/PMC1874340/

(7) The effects of acutely administered 3,4- methylenedioxymethamphetamine on spontaneous brain function in healthy volunteers
Robin Carhart-Harris, et al
International Journal of Neuropsychopharmacology 2013
10.1017/S1461145713001405
Biological Psychiatry 2015 Oct 15;78(8):554-62
http://www.ncbi.nlm.nih.gov/pubmed/24495461

(8) The entropic brain: a theory of conscious states informed by neuroimaging research with psychedelic drugs
Robin L. Carhart-Harris, et al
Frontiers in Human Neuroscience 2014; 8: 20.
http://www.ncbi.nlm.nih.gov/pmc/articles/PMC3909994/

(9) Electroencephalogram signatures of loss and recovery of consciousness from propofol
Patrick Purdon, et al
Proceedings of the National Academy of Sciences vol. 110 no. 12, E1142-E1151
http://www.pnas.org/content/110/12/E1142.full

(10) Thalamocortical model for a propofol-induced alpha-rhythm associated with loss of consciousness
ShiNung Ching, et al
Proceedings of the National Academy of Sciences vol. 107 no. 52, 22665-22670
http://www.pnas.org/content/107/52/22665.abstract

(11) Electroencephalographic Correlates of Vasovagal Syncope Induced by Head-Up Tilt Testing
Fabrizio Ammirati, et al
Stroke 1998 Nov;29(11):2347-51
http://www.ncbi.nlm.nih.gov/pubmed/9804646

(12) Changes in EEG During Graded Exercise on a Recumbent Cycle Ergometer
Stephen P. Bailey, et al
Journal of Sports Science and Medicine (2008) 7, 505 - 511
http://www.ncbi.nlm.nih.gov/pmc/articles/PMC3761919/

(13) Changes in brain cortical activity measured by EEG are related to individual exercise preferences.
S. Schneider, et al
Journal of Physiological Behavior 2009 Oct 19;98(4):447-52.
http://www.ncbi.nlm.nih.gov/pubmed/19643120

(14) Meditation-related activations are modulated by practices needed to obtain it
Barbara Tomosino, et al
Frontiers in Human Neuroscience v.6; 2012 PMC 3539725
http://www.ncbi.nlm.nih.gov/pmc/articles/PMC3539725/

(15) Contribution of dopamine D1 and D2 receptors to amygdala activity in humans
Hidehiko Takahashi, et al
The Journal of Neuroscience, February 24, 2010 • 30(8):3043–3047

(16) Mining the posterior cingulate: segregation between memory and pain components.
Nielsen FA, Balslev D, Hansen LK.
Neuroimage: 2005 Sep; 27(3): 520-32.
https://www.ncbi.nlm.nih.gov/pubmed/15946864

(17) Neurofeedback from the Posterior Cingulate Cortex as a Mental Mirror for Meditation.
Remko van Lutterveld and Judson Brewer
Biofeedback: Fall 2015, Vol. 43, No. 3, pp. 117-120.
http://dx.doi.org/10.5298/1081-5937-43.3.05

(18) Effortless awareness: using real time neurofeedback to investigate correlates of posterior cingulate cortex activity in meditators' self-report
Kathleen A. Garrison, et al 2013
Frontiers in Human Neuroscience v.7; 2013 Aug 6   PMCID: PMC3734786
https://www.ncbi.nlm.nih.gov/pmc/articles/PMC3734786/

(19) Meditation leads to reduced default mode network activity beyond an active task
Kathleen A. Garrison, et al 2015
Cognitive, Affective, & Behavioral Neuroscience
September 2015, Volume 15, Issue 3, pp 712–720
https://www.ncbi.nlm.nih.gov/pmc/articles/PMC4529365/

(20) Dynamical Properties of BOLD Activity from the Ventral Posteromedial Cortex Associated with Meditation and Attentional Skills
Giuseppe Pagnoni, 2012
Journal of Neuroscience 11 April 2012, 32 (15) 5242-5249;
http://www.jneurosci.org/content/32/15/5242.full

(21) Impact of mindfulness on the neural responses to emotional pictures in experienced and beginner meditators
Véronique A. Taylor et al, 2011
NeuroImage, Volume 57, Issue 4, 15 August 2011, Pages 1524-1533
http://www.sciencedirect.com/science/article/pii/S1053811911006070

(22) What about the "Self" is Processed in the Posterior Cingulate Cortex?
Judson A. Brewer, Kathleen A. Garrison, and Susan Whitfield-Gabrieli, 2013
Frontiers of Human Neuroscience, 2013 Oct. 2; 7:647
https://www.ncbi.nlm.nih.gov/pmc/articles/PMC3788347/

(23) Source-space EEG neurofeedback links subjective experience with brain activity during effortless awareness meditation
Remko van Lutterveld et al, 2017
Neuroimage, 2017 May 1;151:117-127. doi: 10.1016/
https://www.ncbi.nlm.nih.gov/pubmed/26921712

(24) Diffusion markers of dendritic density and arborization in gray matter predict differences in intelligence
Erhan Genc, Christoph Fraenz, et al
Nature Communications, 2018, volume 9, Article number 1905
doi:10.1038/s41467-018-04268-8
http://www.nature.com/articles/s41467-018-04268-8
https://www.ncbi.nlm.nih.gov/pmc/articles/PMC5954098/

(25) Cortical glucose metabolic-rate correlates of abstract reasoning and attention studied with positron emission tomography.
Haier RJ, et al,
Intelligence. 1988; 12:199–217.
doi: 10.1016/0160-2896(88)90016-5
https://doi.org/10.1016/0160-2896(88)90016-5

(26) Two dose investigation of the 5-HT-agonist psilocybin on relative and global cerebral blood flow
Candace R. Lewis, Katrin H.Preller, et al
NeuroImage Volume 159, 1 October 2017, Pages 70-78
https://doi.org/10.1016/j.neuroimage.2017.07.020

(27) Neural Correlates of Personalized Spiritual Experiences, Lisa Miller, Iris Balodis et al, *Cerebral Cortex*, May 29, 2018, https://doi.org/10.1093/cercor/bhy102

(31) *Psychedelic Information Theory,* James L. Kent, Amazon, 2010, Chapter 8.

(32) University of North Carolina UNC Health Talk: This is LSD Attached to a Brain Cell Serotonin Receptor, January 26, 2017.
https://healthtalk.unchealthcare.org/this-is-lsd-attached-to-a-brain-cell-serotonin-receptor/   Article cites:

(33) Cell Press (January 26, 2017). *"Structure of LSD and its receptor explains its potency," ScienceDaily.*
https://www.sciencedaily.com/releases/2017/01/170126132541.htm and:

(34) Wacker D, Wang S, McCorvy JD, Betz RM, Venkatakrishnan AJ, Levit A, et al. (January 2017). *"Crystal Structure of an LSD-Bound Human Serotonin Receptor". Cell.* 168 (3): 377–389.e12. doi:10.1016/j.cell.2016.12.033. PMC 5289311. PMID 28129538.
https://www.ncbi.nlm.nih.gov/pmc/articles/PMC5289311/

(35) *Lysergic acid diethylamide*, University of Bristol Chemistry Department article.
http://www.chm.bris.ac.uk/motm/serotonin/LSD.HTM

(36) *Lysergic acid diethylamide*, Wikipedia article.
https://en.wikipedia.org/wiki/Lysergic_acid_diethylamide

(37) Nichols DE (February 2004). "Hallucinogens". Pharmacology & Therapeutics. 101 (2): 131–81. doi:10.1016/j.pharmthera.2003.11.002. PMID 14761703. Archived from the original on February 4, 2016.

Similar articles in Pub Med

- Mystical-type experiences occasioned by psilocybin mediate the attribution of personal meaning and spiritual significance 14 months later.[J Psychopharmacol. 2008]
- Psilocybin occasioned mystical-type experiences: immediate and persisting dose-related effects.[Psychopharmacology (Berl). 2011]
- Validation of the revised Mystical Experience Questionnaire in experimental sessions with psilocybin.[J Psychopharmacol. 2015]
- Classic Hallucinogens and Mystical Experiences: Phenomenology and Neural Correlates.[Curr Top Behav Neurosci. 2018]

Notes:

The experiments cited here were conducted by reputable scientists at distinguished universities who followed rigorous scientific methods. There is every reason to believe their findings are valid, and that the equations which explain them are also true.

That having been said, we need to recognize three things about the neuroscience field.

1. As mentioned in the Neurogenetics chapter, neuroscience is a relatively young field compared to other branches of science. Scientists have only scratched the surface, and much more remains unknown in the field than is currently known. There is still no unanimous consensus among neuroscientists on the functions of some physical regains of the brain or how its cognitive processes work.

2. This nascent science is attempting to decipher the most complex system in the universe. The brain is orders of magnitude more challenging to understand than organic chemistry or the laws of planetary motion.

3. Neuroscientists are working with primitive tools. Despite the phenomenal rate of scientific progress in the 21st Century, we have not seen any major advances in the field of neuroimaging. The methods in use today are all based on 20th Century technologies, and EEG is almost a hundred years old. These measuring instruments lack the depth of penetration and granularity of vision to provide the level of detail scientists need. Conducting neuroimaging experiments with today's equipment has been compared to trying to study molecules with a magnifying glass.

In short, we have a young science studying the most complex system in the known universe with primitive tools. There is a long way to go, and the field is continuously evolving. Neuroscientists are likely to disagree about many topics for years to come, including experiments and interpretations of results. It will be up to future generations of scientists to expand the laws of cognitive physics established here and refine our understanding of the electrodynamics of the brainwave field.

Cognitive Engineering

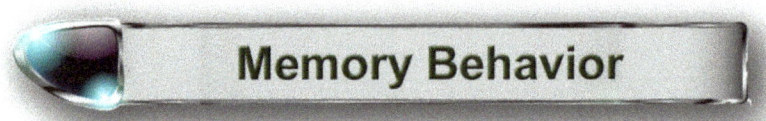
Memory Behavior

● Introduction

This reference chapter explains 14 principles of macro memory behavior, as summarized on the facing chart.

A. Memory System Overview shows the relationship between perception, recording, activation and recall.

B. Applied awareness points power memory waves.

C. Memory waves have 4 types and 3 behaviors

D. Memory records have 2 types and various sizes.

E. Memory recording captures and stores perceptions.

F. Activation and deactivation overview summarizes the 9 steps of memory activation and deactivation.

G. Memory activation is a process where inert memory records are activated by memory waves with similar frequencies.

H. Memory deactivation is a 5 step process.

I. Memory recall causes memory waves to collapse.

J. Memory reconsolidation is a 13 step process which explains how memories can be changed.

K. Memory charge measures the power of memory particles and memory records.

L. Memory activation vectors describe how related memories can activate each other.

M. Memory trees are collections of related memories which are clustered like leaves on memory wave frequency trees.

N. Root anchors are the connection points between memory trees and 16 mental energy centers.

- (A) Memory System Overview
- (B) Applied Awareness Points
- (C) Memory Waves
- (D) Memory Records
- (E) Memory Recording
- (F) Activation and Deactivation Overview
- (G) Memory Activation
- (H) Memory Deactivation
- (I) Memory Recall
- (J) Memory Reconsolidation
- (K) Memory Charge
- (L) Memory Activation Vectors
- (M) Memory Trees
- (N) Root Anchors

Cognitive Engineering

 Memory System Overview

Fig. 15 illustrates the four types of macro memory behavior, which are explained in the following pages.

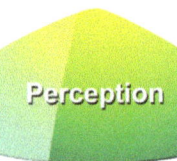

① *Incoming perception:* Neuron pulses arising from an incarnated being's neurology generate perception brainwaves (voltage) which uncondense into space waves. These waves, in turn, uncondense into perception memory waves, which can then enter awareness. A perception memory wave is a higher-order non-physical replica of a brainwave pattern.

② *Experience recording:* A conscious being records its experience in memory waves, which may remain active for a short or long period of time. Eventually, the waves collapse into inert memory particles contained in memory records.

③ *Memory activation:* Active memory waves can transport applied awareness points back into inert memory particles and re-animate them into active memory particles. The activated memory particle emits a replica of the memory wave it originally recorded.

If the being is incarnated, this memory wave condenses through space into voltage waves, which overlay themselves on neurology. Behavior memory waves condense into brainwave patterns which inform neurology to conduct a remembered action, such as turning a doorknob or speaking a word.

④ *Memory recall:* A re-activated memory wave may potentially enter consciousness and be recalled. If it does, the wave collapses into pure information in consciousness.

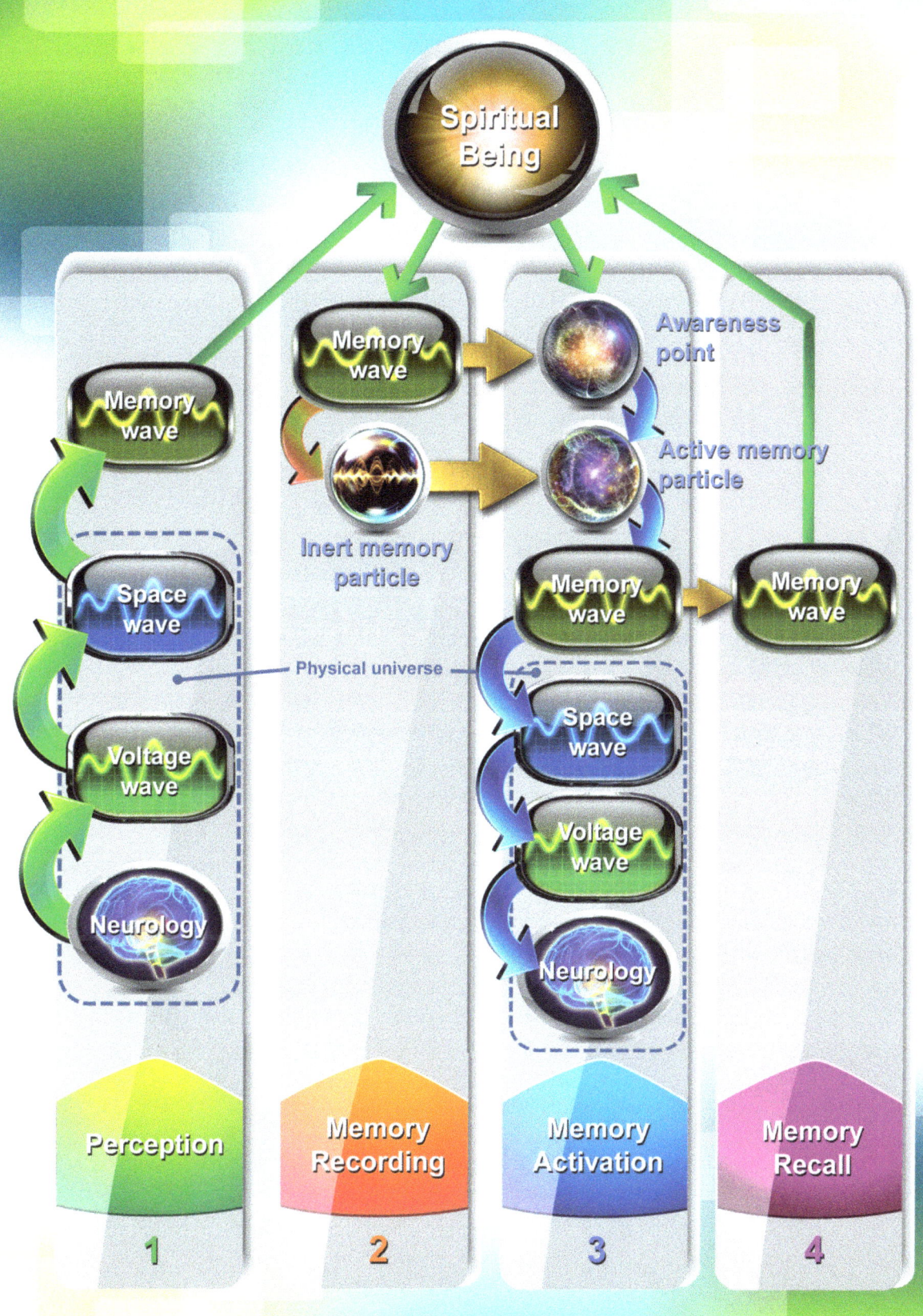

– Fig. 15 –

Cognitive Engineering

B) Applied Awareness Points

Free awareness points become applied awareness points when they assume a form. When a memory record is activated, applied awareness points assume the form of the memory particles in the record. The applied awareness points are the power source which provides the energy that generates the transmission of a memory wave.

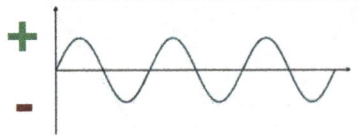

The form which the awareness point assumes has polarity. Memory particles are recordings of memory waves. Waves have polarity. As shown in the diagram, the peaks in the waves are positive and the valleys are negative.

The awareness point itself has no inherent polarity, but when it activates a memory, it animates a form which does. In so doing, it assumes the form's polarity.

The applied awareness point changes polarity at a rate governed by the frequency signature of the memory wave recorded in the memory particle.

For example, if the memory wave has a frequency of 20 Hertz, the applied awareness point changes state 20 times per second. If the memory wave has a frequency of 35 Hertz, the applied awareness point changes state 35 times per second.

These changes in state are called *oscillations* and they generate memory waves. An oscillating applied awareness point generates a memory wave. The key attribute of a wave is oscillation.

A similar principle operates in the physical universe. In electrodynamics, an oscillating electric field produces an oscillating electromagnetic field (Ampere's Law).

Oscillating awareness points behave like electrons in an oscillating electric field to supply the power which generates wave transmissions.

Cognitive Engineering

 Memory Waves

● Types
The four classes of memory behavior shown in Fig. 7 are powered by four distinct types of memory waves:

① *Incoming perception from neurology:* Perception memory waves are higher-order spiritual memory replicas of perception brainwave patterns.

② *Experience recording:* Memory waves are emitted by consciousness.

③ *Memory record activation:* An active memory record emits a replica of the memory wave it originally recorded

④ *Memory recall:* Activated memory waves may potentially be recalled.

Examples of memory waves include:
- The motor neuron firing patterns for opening a doorknob
- Meanings of words spoken, written or thought
- Face recognition
- Object recognition

● Behaviors
Memory waves are higher-order octaves of electromagnetic wave patterns, carrying information in frequency and amplitude-encoded signals. A memory wave has three kinds of behaviors:
- A. It can be recalled.
- B. It can remain actively unconscious.
- C. It can deactivate and become pre-unconscious (inert).

A. Conscious Recall: Memory waves can enter consciousness. When the memory wave is consciously recalled, it collapses into pure information and understanding in the conscious being. (Consciousness has no wavelength.) The old wave itself ceases to exist. The information in consciousness is recorded in new memory waves. If the information is altered while in consciousness, the changes will be reflected in the new memory waves.

B. Unconscious Activity: All memory waves are unconscious. As long as they remain in force, if the being is incarnated, they will condense and propagate through successive layers of space, electromagnetic waves, and neurology. These space waves and electromagnetic waves are the physical universe manifestations of life. The awareness point (consciousness) and the memory wave (unconsciousness) are the sources of life.

C. Unconscious Deactivation: Memory waves can remain active for centuries. Eventually, they deactivate on their own accord, condensing into inert memory particles, and releasing the living awareness points which power them back into free consciousness.

Behavior Examples

1. Activation: Whenever someone opens a door, they activate and utilize their memory of how to execute the motion. Their unconscious mind performs the activity for them without them having to be consciously aware of it.

2. Deactivation: When we are finished using unconscious memories, they are dismissed from active unconsciousness. Memories pass in and out of active unconsciousness continuously without ever being consciously recalled.

3. Recall: Active memories have the potential for being consciously recalled. A memory must be active for it to be recalled.

D Memory Records

1. Definition

A memory record is a conscious being's recording of non-physical memory waves. There is one memory record per unit of time.

Memory records contain different memory particles which are associated by time. The memory record contains recordings of all perceptions – visual, auditory, spatial orientation, temperature, sensation, taste, smell, tactile and so on – as well as all internal cognitive activity such as thoughts, emotions and attitudes. Each perception is encoded in and carried by a separate memory wave.

The memory record is like a template, blueprint or film frame for re-generating the memory waves it originally recorded. The memory record has no life of its own. Only the awareness point and the memory wave are alive.

2. Types

Perception: The *memory perception record* is a recording of all memory waves resulting from brainwaves generated by perceptions in one unit of time. It comprises memory particles which each store a recording of one individual memory wave.

Behavior: The *memory behavior record* is a recording of all memory waves which generate brainwaves in one unit of time. It comprises memory particles which each contain a recording of one individual memory wave.

3. Topography

The vast majority of memory records are inert. Only a tiny fraction of memory records are ever active at one time. Inert memory is the overarching component of mental geography, with active memory just a thin veneer on the surface.

If a person has had a few hundred lifetimes, their current life occupies less than one percent of their memory records.

– Fig. 16 –

E. Memory Recording

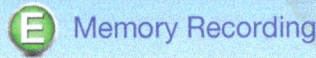

1. Perception

Memory must be recorded before it can be activated, deactivated or recalled, so examining memory recording is the starting point for understanding the rest of memory behavior.

Something must be perceived before it can be remembered, so Figure 17 illustrates how incoming perceptions are converted into memory waves.

Referring to Fig. 17, inputs from the body's perceptual systems 1 travel to the brain 2, generating voltage brainwave 3, which propagates upwards and uncondenses into space wave 4 (as covered earlier in the discussion of propagation).

Space wave 4 strikes a spirit/space membrane 5 (as discussed under propagation layer 3 – space waves), causing free awareness points 6 near the membrane to oscillate in a resonant frequency. The stationary free awareness points 6 become oscillating applied awareness points 7, whose fluctuations produce memory waves 8.

Memory waves 8 have the potential for being perceived by the conscious portion of the being, as indicated by conscious perception 9. They are always "perceived" by the unconscious portion of the being.

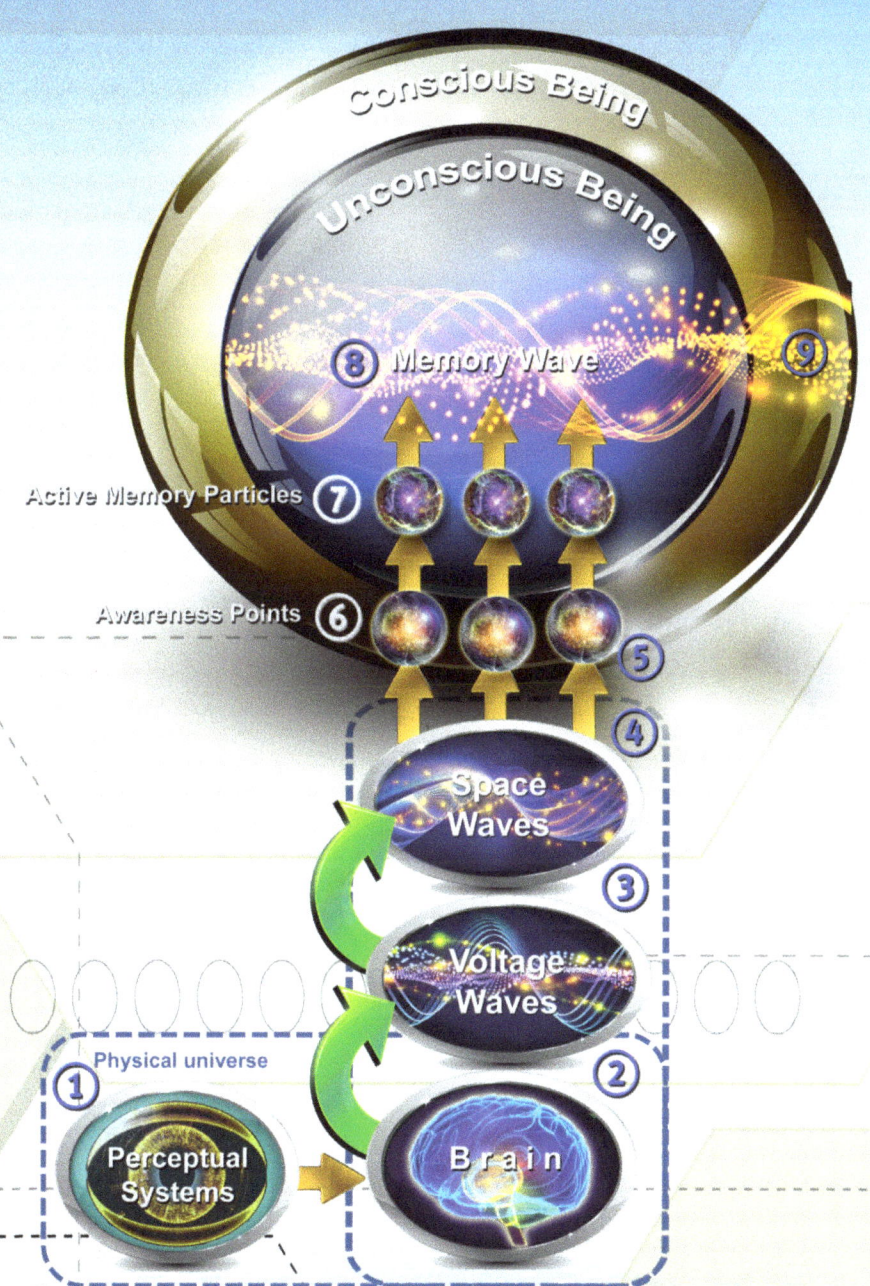

– Fig. 17 –

2. Recording

Once perceptions have been converted into memory waves, they can be recorded. Figure 18 shows how memory waves are converted into inert memory particle recordings.

Unconscious memory wave 1 may remain in force for seconds, days, years or centuries. Eventually, it deactivates, and the applied awareness points 2 stop oscillating and change state back into free conscious awareness points 3. Memory wave 1 collapses into inert memory particles 4 which comprise inert memory record 5.

The deactivated memory record can remain dormant indefinitely, or become activated again. If it becomes activated, it has the potential for being recalled.

Memory must be recorded before it can be activated, deactivated or recalled.

To summarize, memory recording is the result of awareness points being applied to create memory waves which record information in time. The awareness points begin to oscillate, and the oscillations generate memory waves. Eventually, the waves condense into inert particles, and the awareness points return to the being's reservoir of free conscious awareness.

Memory Recording

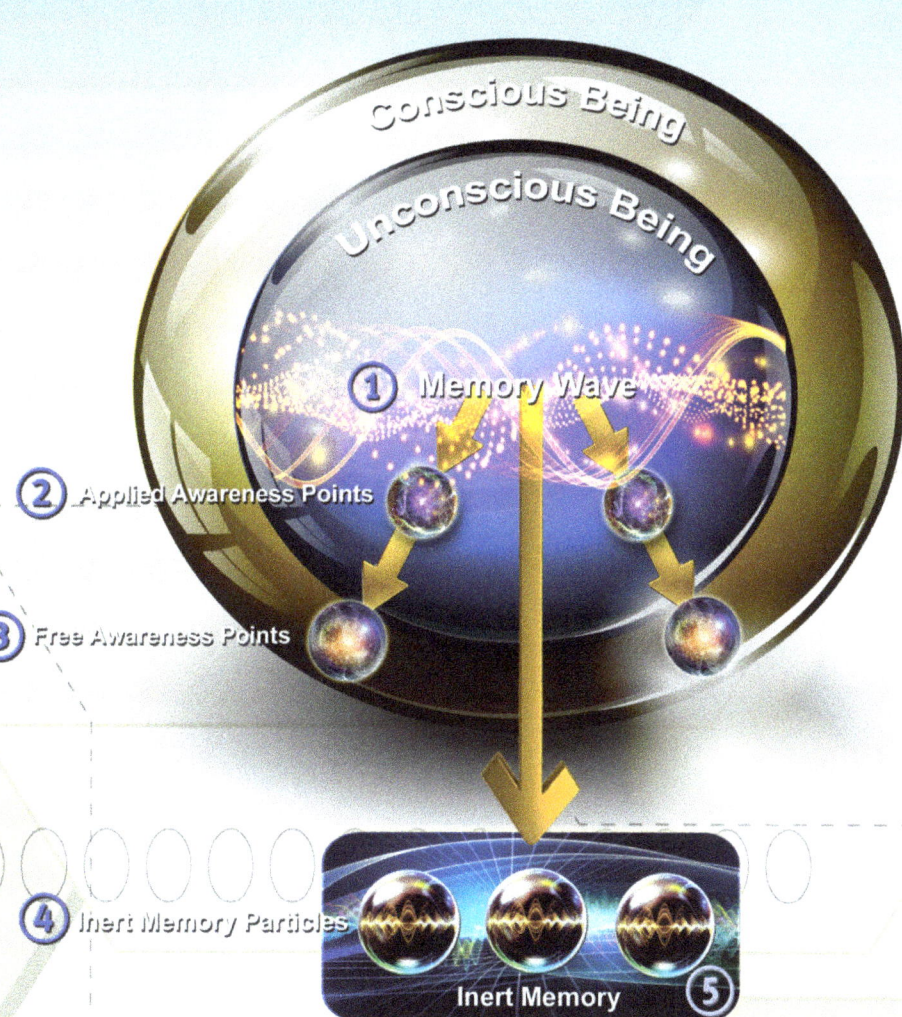

– Fig. 18 –

Memory Activation and Deactivation Overview

The memory activation and deactivation cycle comprises 8 steps which are depicted in Fig. 19 and described below.

Memory Activation

① The cycle begins with Awareness Point 1.

② Awareness Point 1 is vibrated by incoming Perception A.

③ Awareness Point 1 changes its state into Perception Memory Wave A.

④ If Perception Memory Wave A has sufficient amplitude, it strikes Associated Inert Memory Particle B, which holds a recording of a similar memory wave. Perception Memory Wave A transports the life energy of Awareness point 1 into Associated Inert Memory Particle B

⑤ Associated Inert Memory Particle B is transformed into Active Memory Particle B.

⑥ Active Memory Particle B emits Activated Memory Wave B, which may remain active for a short or long period of time.

Memory Deactivation

⑦ Eventually, Memory Wave B collapses back into Inert Memory Particle B.

⑧ Inert Memory Particle B releases Awareness Point 1 back into free conscious awareness.

Memory Activation & Deactivation Overview

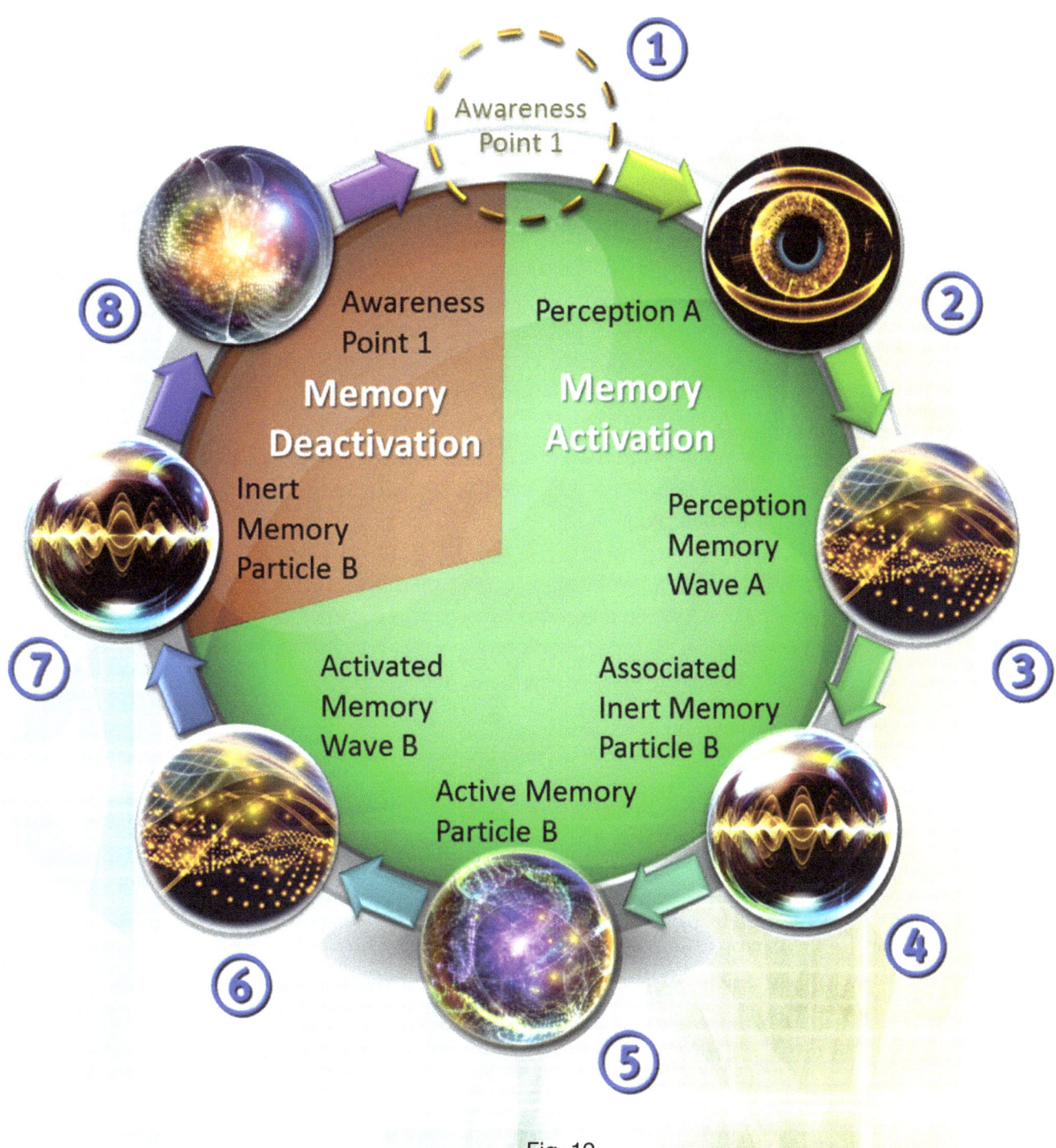

– Fig. 19 –

Cognitive Engineering

G Memory Activation

Referring to Fig. 20, a current perception memory wave 1 (such as a bird image) activates inert memory particle 2 in inert memory record 3, energizing it into active memory particle 4 in active memory record 5. Active memory particle 4 emits memory waves 6 which are replicas of the past memory waves it originally recorded.

Once in a wave state, the memory has the capacity for entering consciousness and being recalled. If the memory is not recalled, it remains unconscious, and the memory waves propagate themselves through physical space and animate neurology if the being is incarnated.

As discussed earlier in the section on propagation, memory waves 6 condense into space waves 7, which condense into electromagnetic waves 8, which, in turn, overlay themselves on neurology 9.

Memory waves 6 can also activate other similar memories. An example is shown here with memory wave 6 activating memory particle 10 in memory record 11. The activated memory particle 10 emits memory wave 12, which condenses into space waves 13 and voltage waves 14 as explained above.

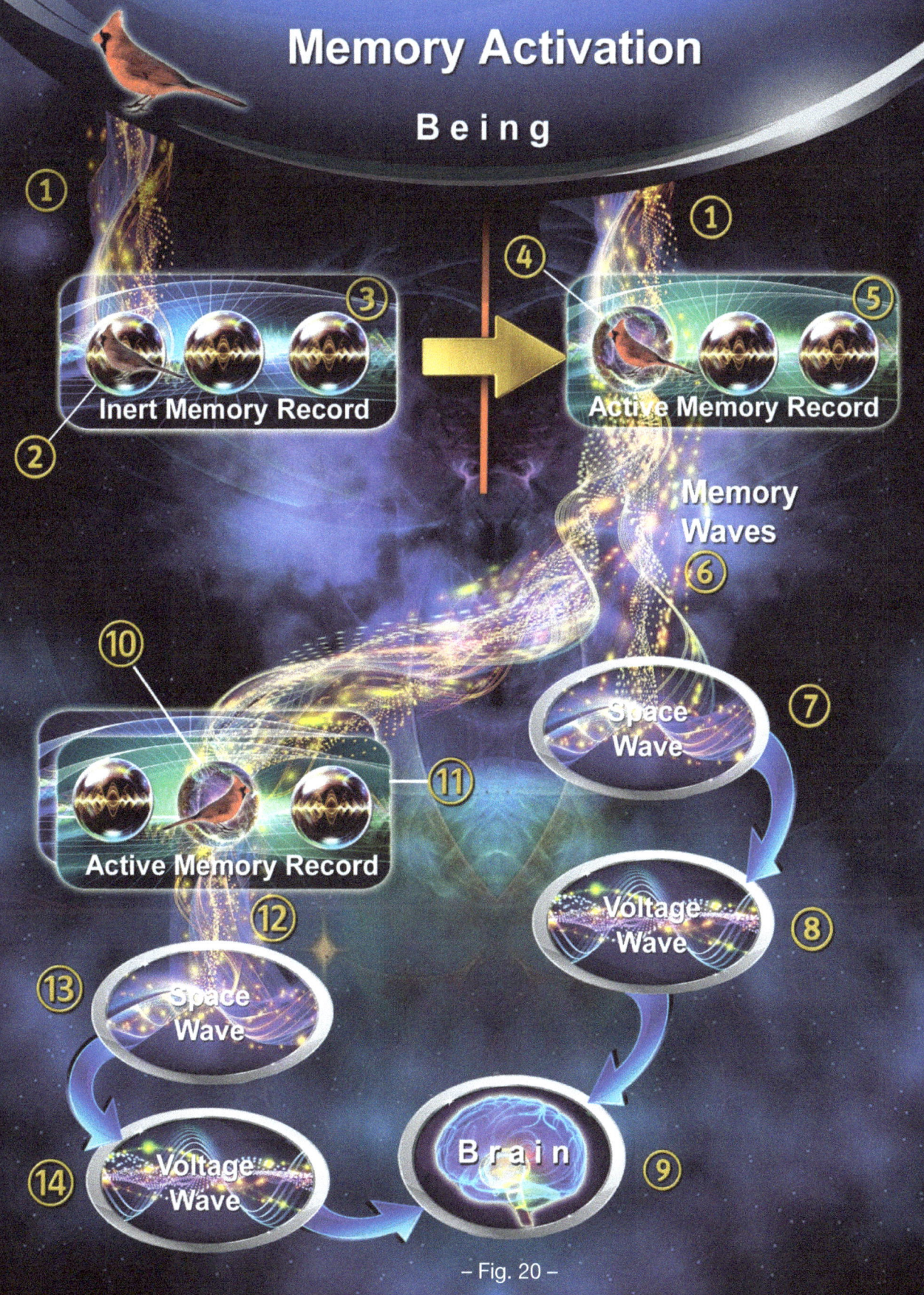
– Fig. 20 –

H Memory Deactivation

Referring to the left panel of Fig. 21 on the facing page, labelled "Active," applied awareness points 1 are animating active memory particles 2 which emit unconscious memory waves 3. Memory waves 3 draw their power from the applied awareness points 1. The memory waves remain active as long as awareness is applied to the memory particles 2.

The right panel of Fig. 12, labelled "Inactive," shows how memory is deactivated when awareness is removed from the memory record. In the right panel, the applied awareness points have left the memory record, returning to the conscious being as free awareness points 4. Memory waves 3 in the left panel have condensed into inert memory particles 5 in inert memory record 6 in the right panel.

The active memory particles become inert, and the memory record becomes inert. The applied awareness points that were the force of the memory waves' projection rejoin the main body of free awareness.

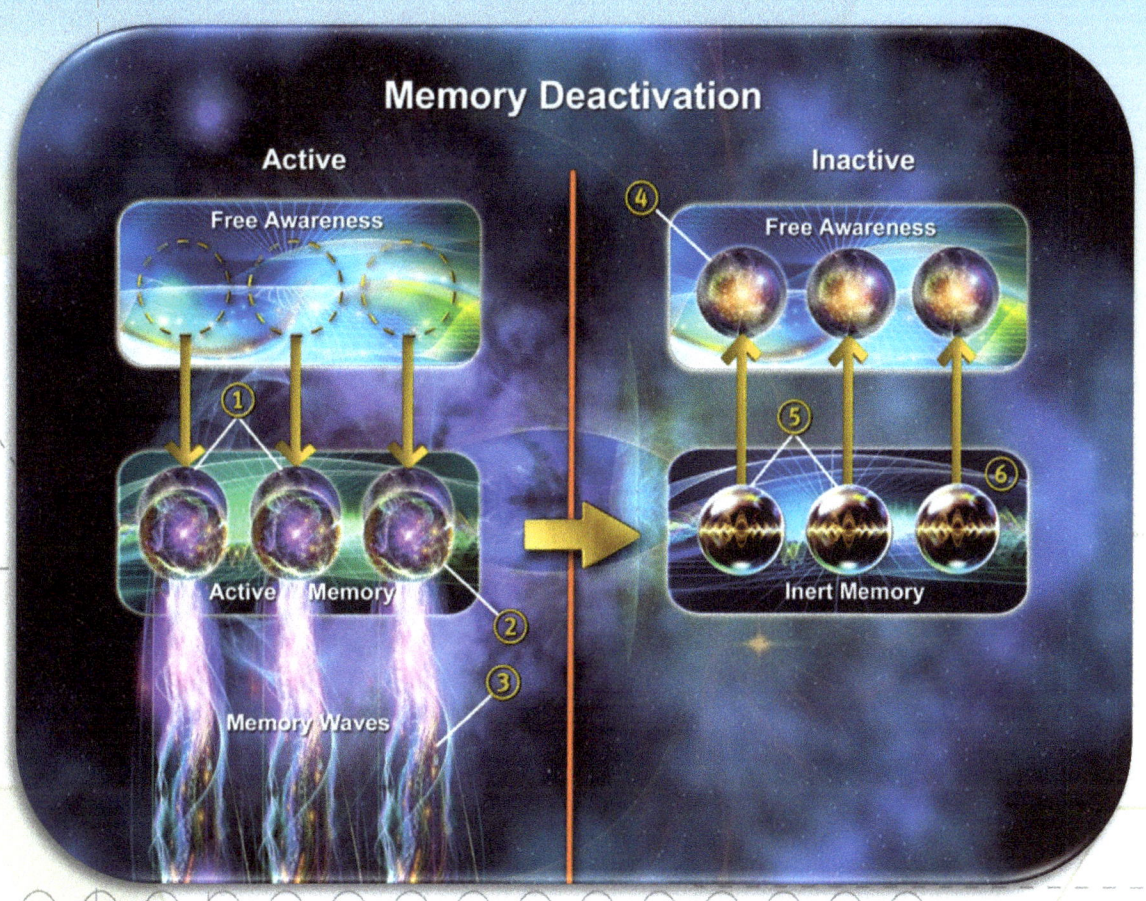

– Fig. 21 –

Cognitive Engineering

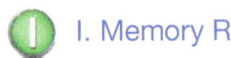
I. Memory Recall

When inert memory particles are activated by applied awareness points, they change state into memory waves which can enter consciousness.

When memory waves are recalled by the conscious being, their wave patterns collapse into pure information. (Conscious awareness has no wavelength.) As shown in Fig. 22, this causes the memory waves and memory record to collapse.

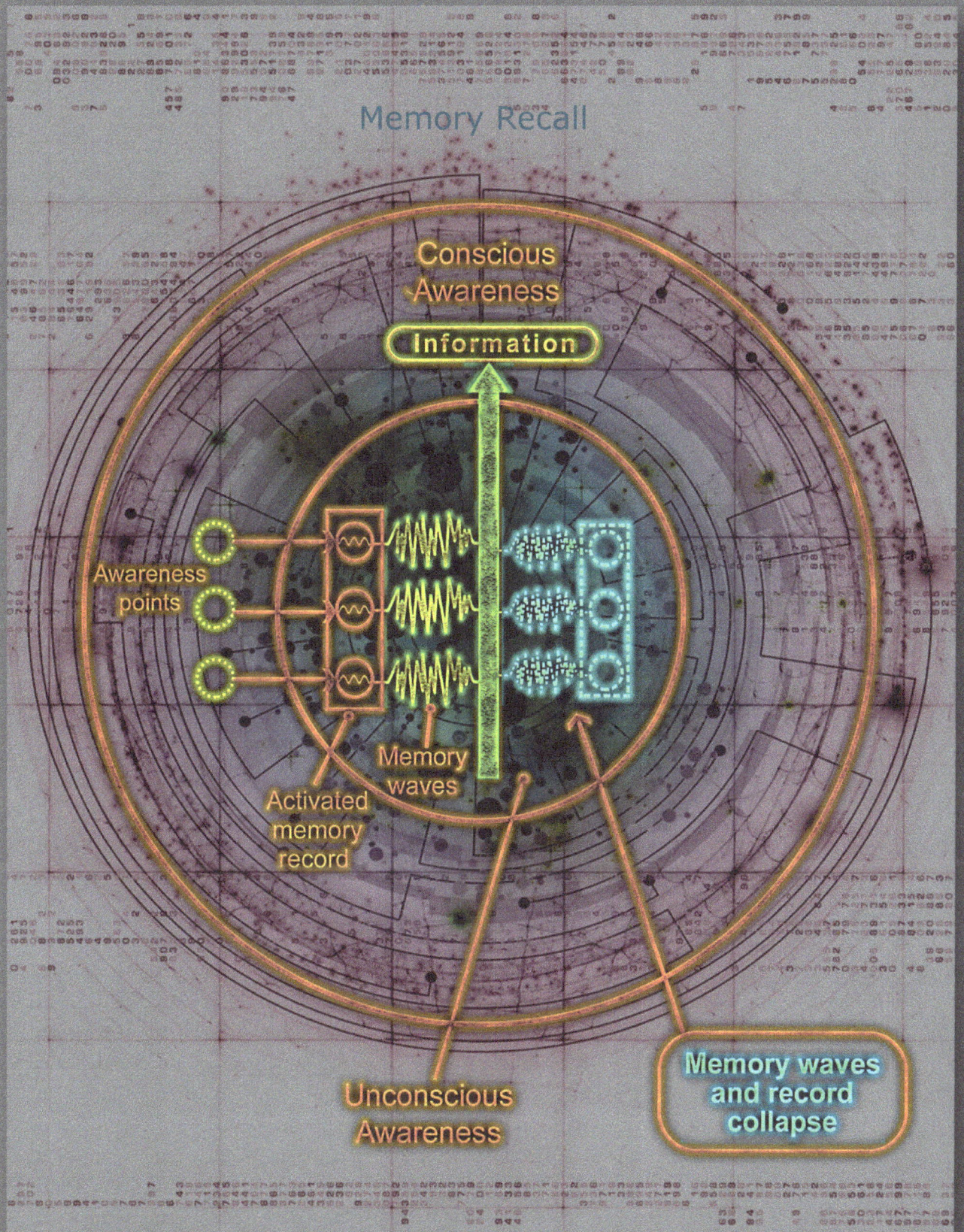

– Fig. 22 –

Cognitive Engineering

 Memory Reconsolidation

Referring to Fig. 23, a person's incoming perception of a cardinal is conveyed by memory wave 2 which is animated by awareness point 1. Seeing one cardinal unconsciously reminds the person of a time they saw two cardinals. Memory wave 2 activates inert memory particle 3, transforming it into active memory particle 4, which emits new memory wave 5, containing both birds.

New memory 5 enters consciousness as recalled memory 6, releasing awareness point 7. Now the memory only exists as pure information in consciousness. Memory particle 4 and memory wave 5 cease to exist, as illustrated by vanishing points 8.

While in consciousness, the memory is altered from left-facing birds 6 to right-facing birds 9. This new experience in consciousness is recorded by memory wave 11, powered by awareness point 10. Eventually, memory wave 11 collapses into inert particle 12, releasing awareness point 13. Inert memory 12 of the right-facing birds has replaced inert memory 3 of the left-facing birds.

When a memory is recalled, the information it contains is transferred into conscious awareness, and the memory itself ceases to exist.

The Propagation Corollary informs us that changes in physical neurology are condensed manifestations of higher-level mental activity in the conscious being. Accordingly, physical-level memories should disappear when their contents are consciously recalled.

The first evidence of this was observed by Dr. Karim Nader in the late nineties at McGill University. Dr. Nader's experiments showed that recalled memories are erased from physical neurology when their contents are transferred into consciousness. He called the process "memory reconsolidation." [16]

---

[16] *Memory Reconsolidation*, edited by Christina Alberini, Elsevier Academic Press 2013, chapters 1 & 2.

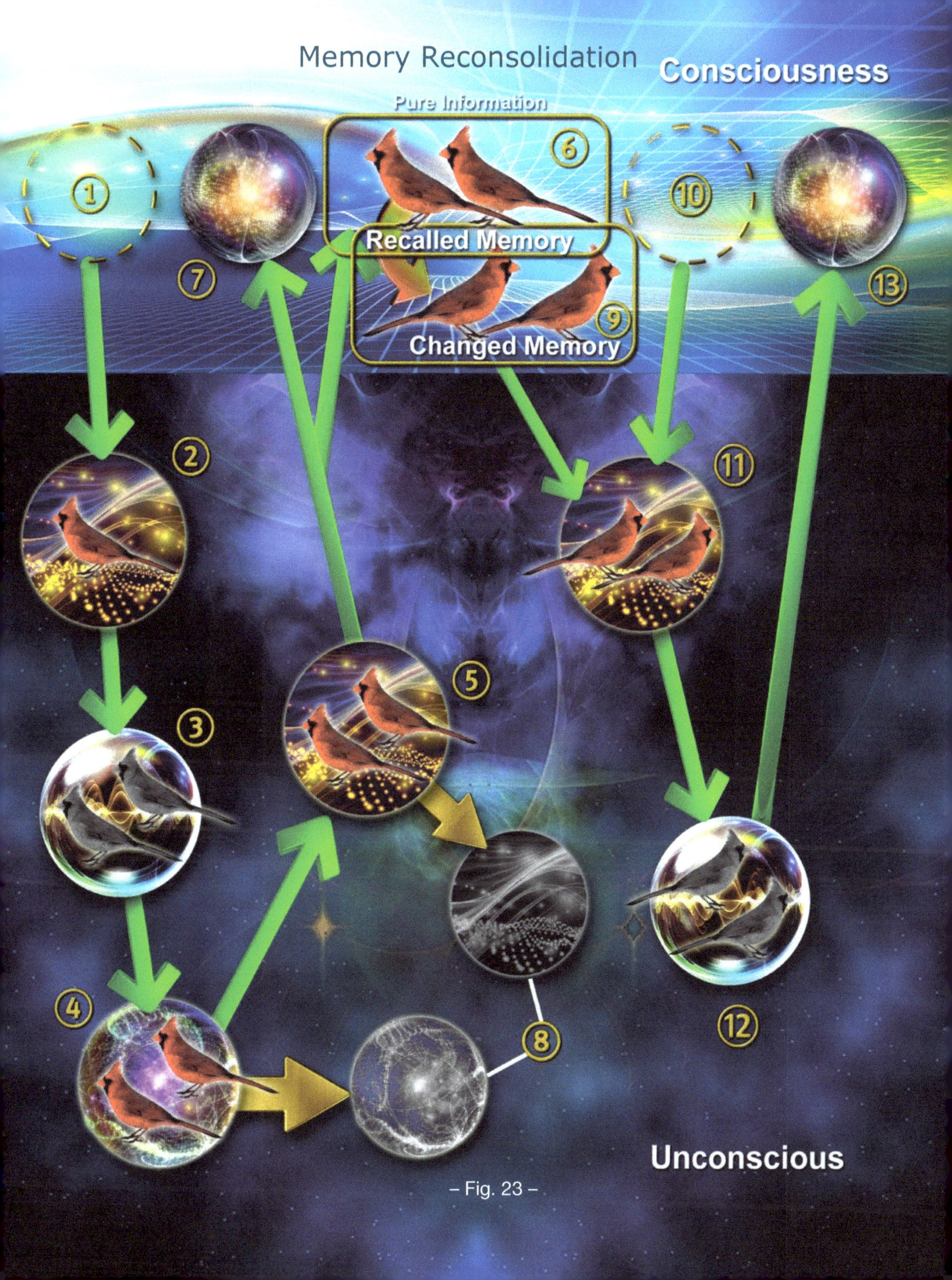
Memory Reconsolidation
Consciousness
Pure Information
Recalled Memory
Changed Memory
Unconscious
1
2
3
4
5
6
7
8
9
10
11
12
13

– Fig. 23 –

Cognitive Engineering

The reason past life regression therapy can alleviate unwanted conditions is that memories can be altered. Regression therapy can change a subject's memories and discharge their force. The new memories are rewritten into unconsciousness, replacing the old ones.

 Memory Charge

a. Memory Power

Some memory waves have lower amplitudes than others. For example, the sensation of jumping into a lake normally has lower energy than the sensation of jumping out of an airplane, because skydiving is more exciting for most people.

When memory waves are recorded, they collapse into memory particles. Larger memory waves form bigger memory particles. As Fig. 24 shows on the facing page, larger memory particles require more life energy to activate than smaller ones.

The amplitude of a reactivated memory wave is proportional to the number of awareness points which are energizing the active memory particle emitting it. In a fully-activated memory, this number will be equal to the number of awareness points which animated the original memory wave.

A memory particle cannot absorb more awareness points than the number carried by the original memory wave it recorded.

The amplitude of a memory wave depends on the number of applied awareness points driving it. High-amplitude memory waves require more awareness points to power them.

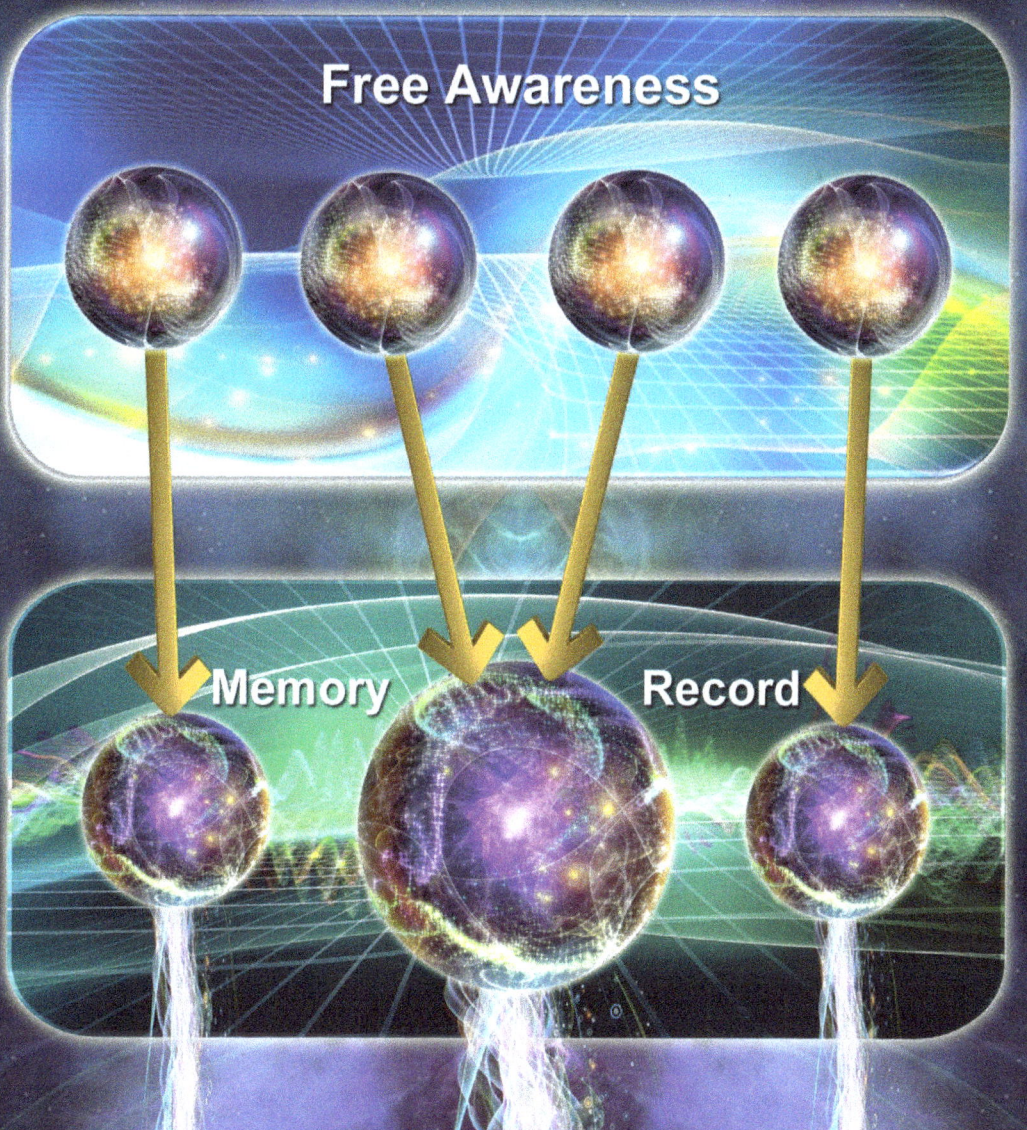

– Fig. 24 –

b. Partial Memory Activation

Memory records can be fully or partially:
        a) activated
        b) deactivated
        c) recalled

Fig. 16 on the facing page shows a partially-activated memory record. Notice the record contains a mixture of inert and active memory particles.

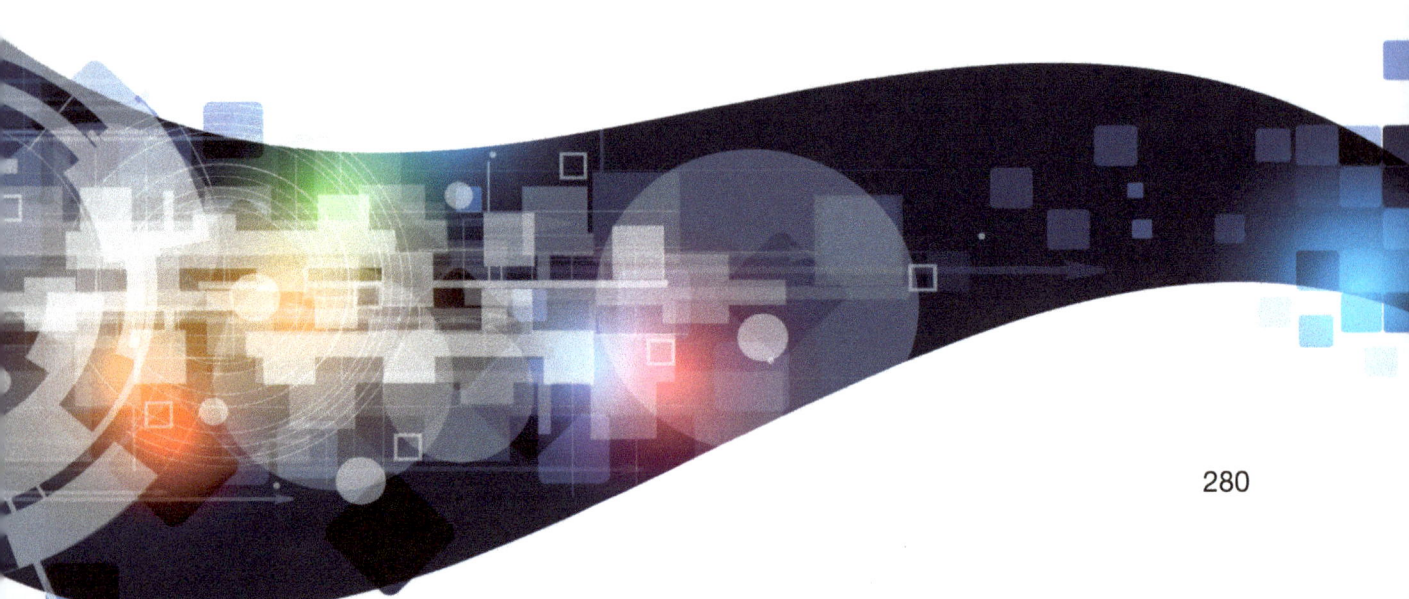

– Fig. 25 –

c. Memory Record Charge

How many awareness points does it take to activate a memory record? The answer depends on the level of mental activity recorded during the snapshot in time the memory represents, including its sensory and emotional amplitude. The number and size of the waves in the memory determine how many awareness points are required to re-generate it. The higher amplitude the brainwaves had when originally recorded, the more awareness points it takes to reproduce them.

The strength of a memory record's wave field is a function of the number of applied awareness points enclosed within the memory record.[17] The outward flow of memory waves (flux density) emanating from a memory record is proportional to the number of applied awareness points in the record.

The amplitude of the memory wave is proportional to the number of awareness points animating the memory. A memory record can be fully or partially activated. If the memory record is only partially activated, it will contain both active and inert memory particles. Inert memories have no measurable charge; they only have "potential." Only active memories have measureable charge.

The strength of a memory wave is determined by the stored potential of the memory record, like the output of a battery is determined by its voltage rating. The stored potential of the memory record governs the number of awareness points required to activate it.

Formulas are given below for calculating the charge of active, inert, and partially-active memory records.

---

[17] This principle has a parallel law in electrical engineering, which states the electric flux on the outside of a sphere is proportional to the number of electrons enclosed within the sphere (Gauss' First Law).

Active Memory Record

The amplitude of an active memory record equals the sum of the amplitudes of its component active memory particles.

Inert Memory Record

The potential amplitude of an inert memory record equals the sum of the potential amplitudes of its component inert memory particles.

Partially-Active Memory Record

The potential amplitude of a partially-active memory record equals the sum of the amplitudes of its active memory particles plus the sum of the potential amplitudes of its inert memory particles.

d. Memory Wave Field Charge

The amplitude of a being's total unconscious memory wave field is proportional to the number of applied awareness points animating memory waves enclosed within the field.

Memory Activation Vectors

Related memories can activate each other in two ways, as shown in the illustration.

Referring to Fig. 26, when memory wave 1 activates memory particle 2, memory particle 2 emits secondary memory wave 3.

If memory wave 1 has sufficient amplitude to activate one or more additional memory particles, it can flow in two directions.

Time vector: Memory wave 1 can activate adjacent memory particles 4 and 5, which emit memory waves 6 and 7. As shown by their red and green colors, these two particles contain different memory wave recordings than memory particle 2 (for example, the image of a cardinal, the sound of its birdsong, the crisp feel of cool morning air).

The three particles in the memory record are associated closely together in time. Time is the underlying medium of a memory record. United by a common fabric of time, memory particles in the same record can influence one another.

Frequency vector: Memory wave 3 can also activate other matched-frequency memory particles located in different memory records, such as memory particle 8. This particle emits memory wave 9, which, in turn, can light up memory particle 10 as shown. Memory particle 10 then emits memory wave 11.

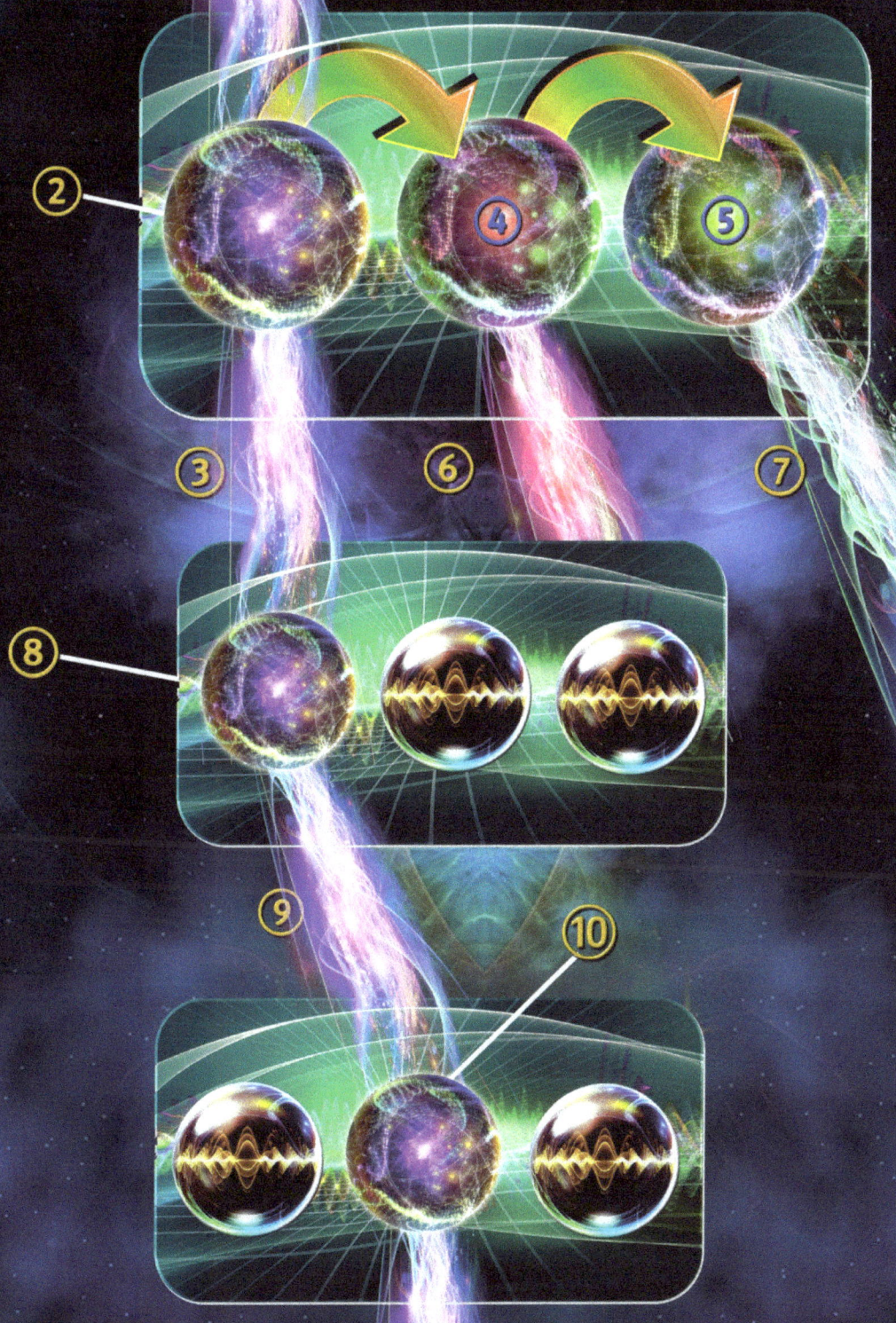

– Fig. 26 –

Cognitive Engineering

Memory Activation Vectors

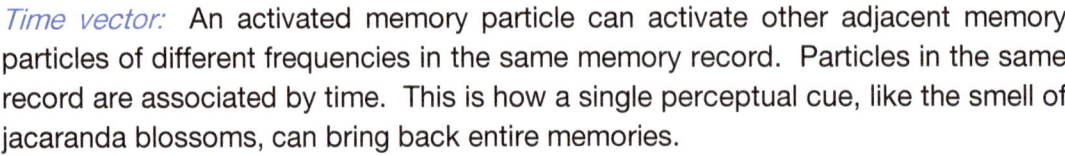

Time vector: An activated memory particle can activate other adjacent memory particles of different frequencies in the same memory record. Particles in the same record are associated by time. This is how a single perceptual cue, like the smell of jacaranda blossoms, can bring back entire memories.

Frequency vector: An activated memory particle can activate other memory particles of the same frequency in different memory records. This explains how one memory can remind us of another (consciously or unconsciously). It is also the central organizing principle of *memory trees*, which are explained on the following pages.

The greater a memory wave's amplitude, the more memory particles it can activate.

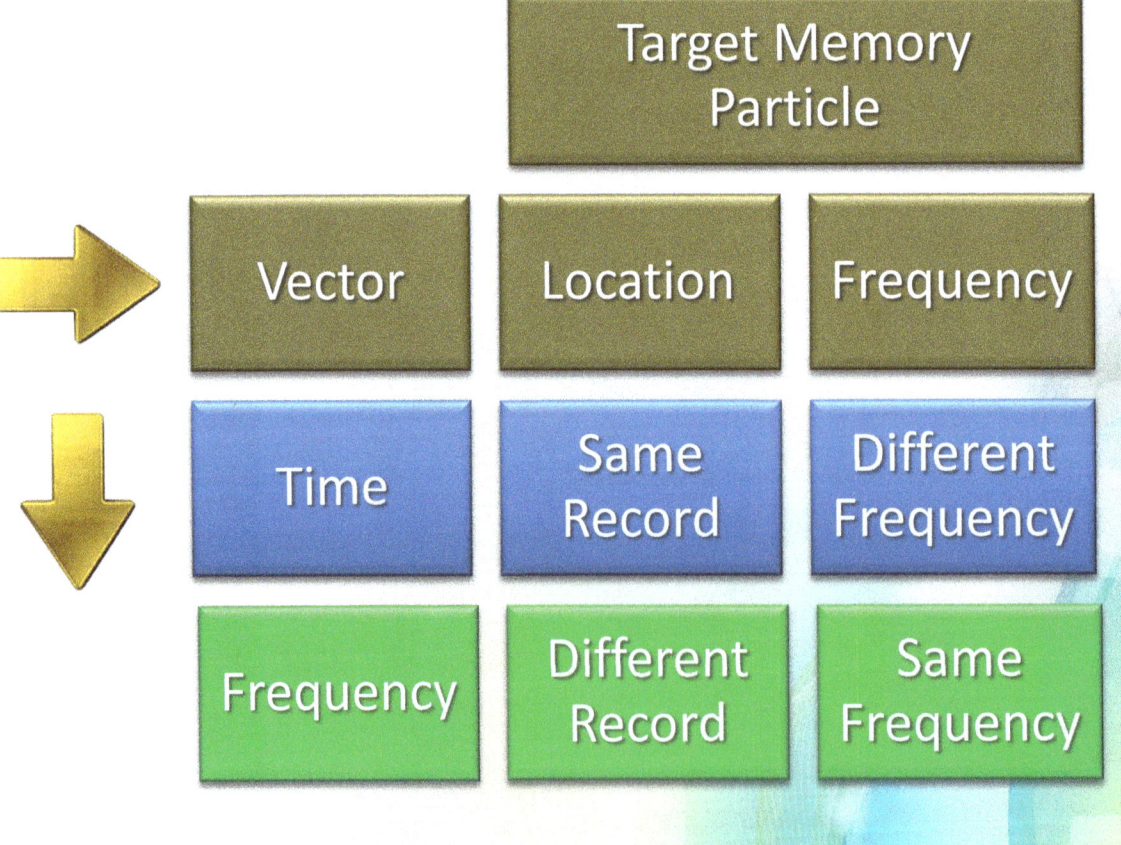

– Fig. 27 –

Memory Trees

Memory trees are the key organizing principle of active unconscious memory associations. A memory tree is a collection of related memory records which all share a common memory wave frequency signature, for example, caramel colored cats. Memories cluster like leaves on memory wave frequency trees.

Memory records are "logical"; they contain physically-different memory particles which are associated by time. *Memory trees* are "physical"; they consist of physically-similar memory records which have at least one memory particle with the same frequency in common. The frequency of the common particle is identical for all memory records on the tree, and it can be thought of as the frequency of the memory tree.

Memory trees are constantly forming and disappearing. They can remain active for a few seconds or many centuries. A person can have thousands of active trees in their unconscious.

Every type of memory can have its own tree. For example, imagine the *pet-a-cat* memory tree is located on a specific band of the memory wave frequency spectrum. Suppose there are 5 "leaves" on the tree representing 5 active memory records. Each of these memory records has an active memory particle containing a memory wave recording of the tactile sensation of *petting a cat*.

The memory records themselves might be of petting an Egyptian Mau, a Russian Blue, a Siamese, a Shorthair and an Abyssinian. Although the records have different contents, they all contain an active memory particle with the same tactile sensation of petting a cat. The active memory particle joins the memory record to the *pet-a-cat* tree like a leaf stem attaches a leaf to a branch.

The conscious being does not have to exert any effort consciously to activate memory particles. They are activated automatically by memory wave frequencies like cell phones are activated by incoming calls.

The first memory record in the tree is called the *anchor*. This memory record anchors the memory tree and connects it to the life energy of the conscious being. It is analogous to physical roots which enable living trees to draw water from the earth.

Fig. 28 on the facing page illustrates an simple memory tree with five memory records. For simplicity, each memory record is shown to contain three memory particles, although actually, a memory record can hold dozens of different memory particles containing thoughts, attitudes, perceptions and sensations. The unique frequencies of these memory particles are symbolized in the diagram by different colors.

A tree of memories that involves jacaranda blossoms will be connected by the flower's scent. The olfactory sensation has a specific frequency, and its memory wave can activate memory particles with that frequency like telephone calls ring cell phones.

The common memory wave of the flower's scent is like the wood of the memory tree, and the memory records of individual experiences are like leaves. Each memory record is connected to the memory tree through a *node particle*, like leaf stems on a living tree are attached to branches by nodes.

The node particle in the memory records on each tree is the memory particle which has the same recorded frequency as the common memory wave of the tree. In this example, the node particles are the parts of each individual memory which contain the jacaranda blossom's scent. (Hence, they are colored lavender in the diagram.)

Referring to Fig. 19, memory wave 1 activates root node memory particle 2 in root anchor memory record 3. Root node 2 activates upstream node particles 4 which connect the remaining memory records to the tree.

Fig. 29 which follows illustrates memory tree clusters.

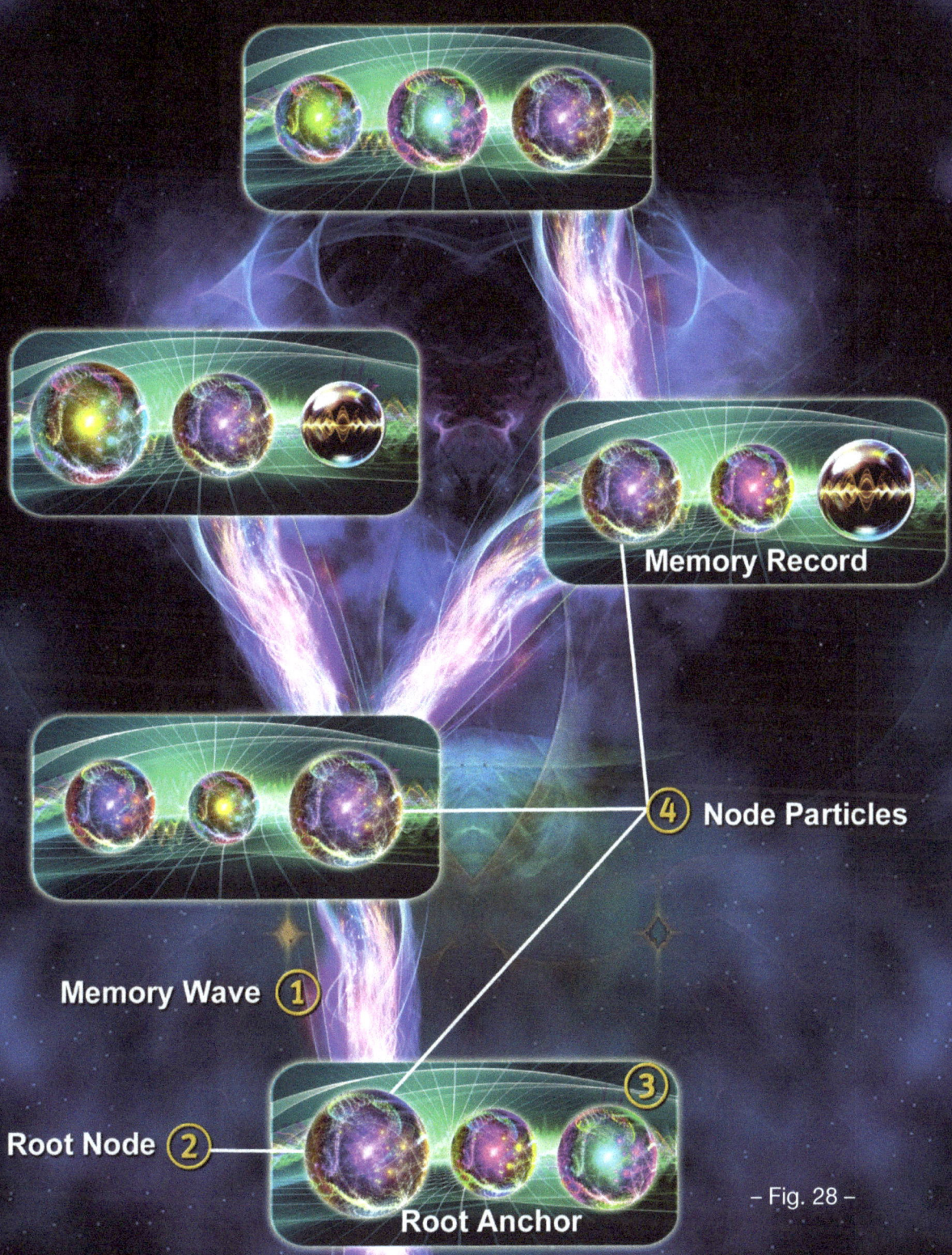

– Fig. 28 –

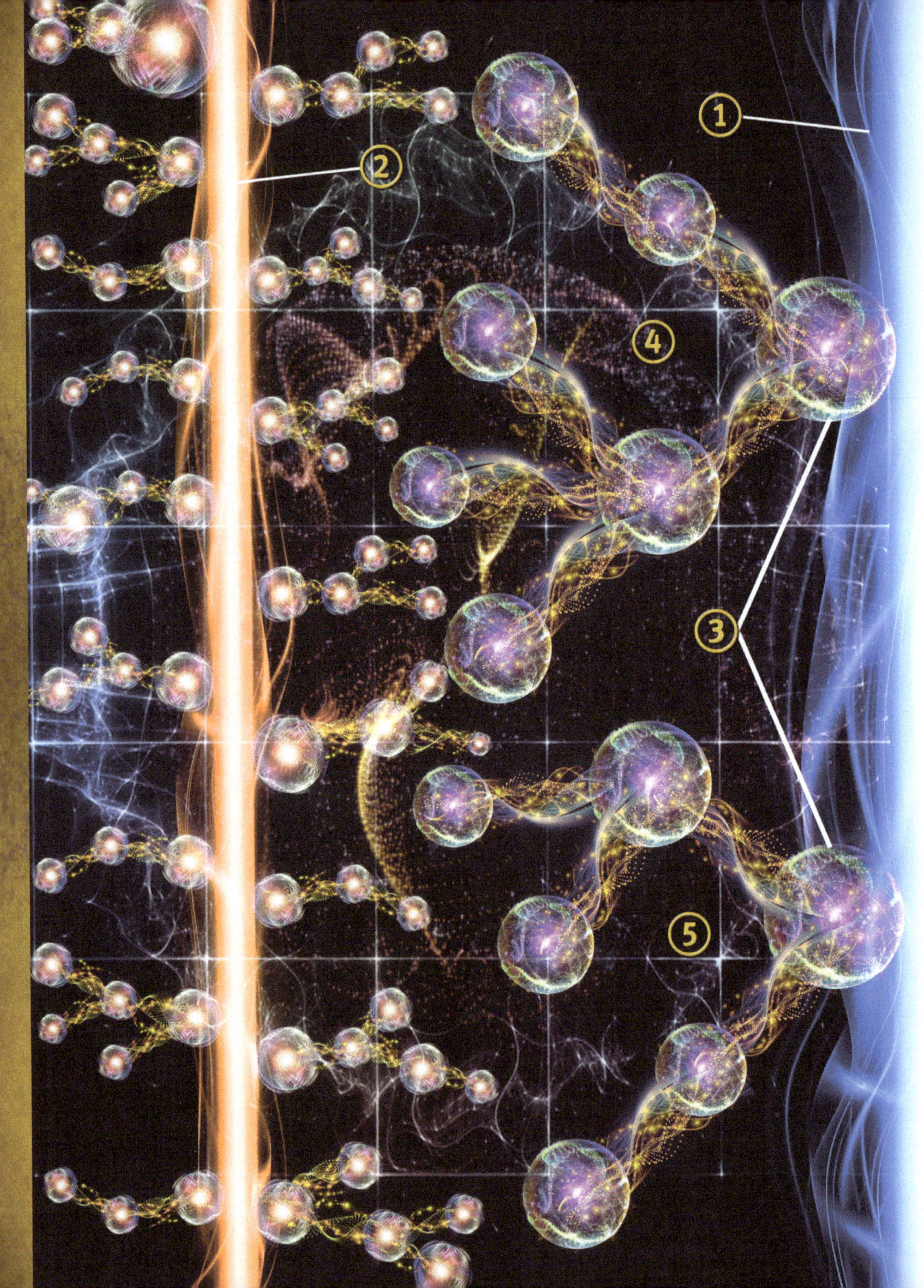

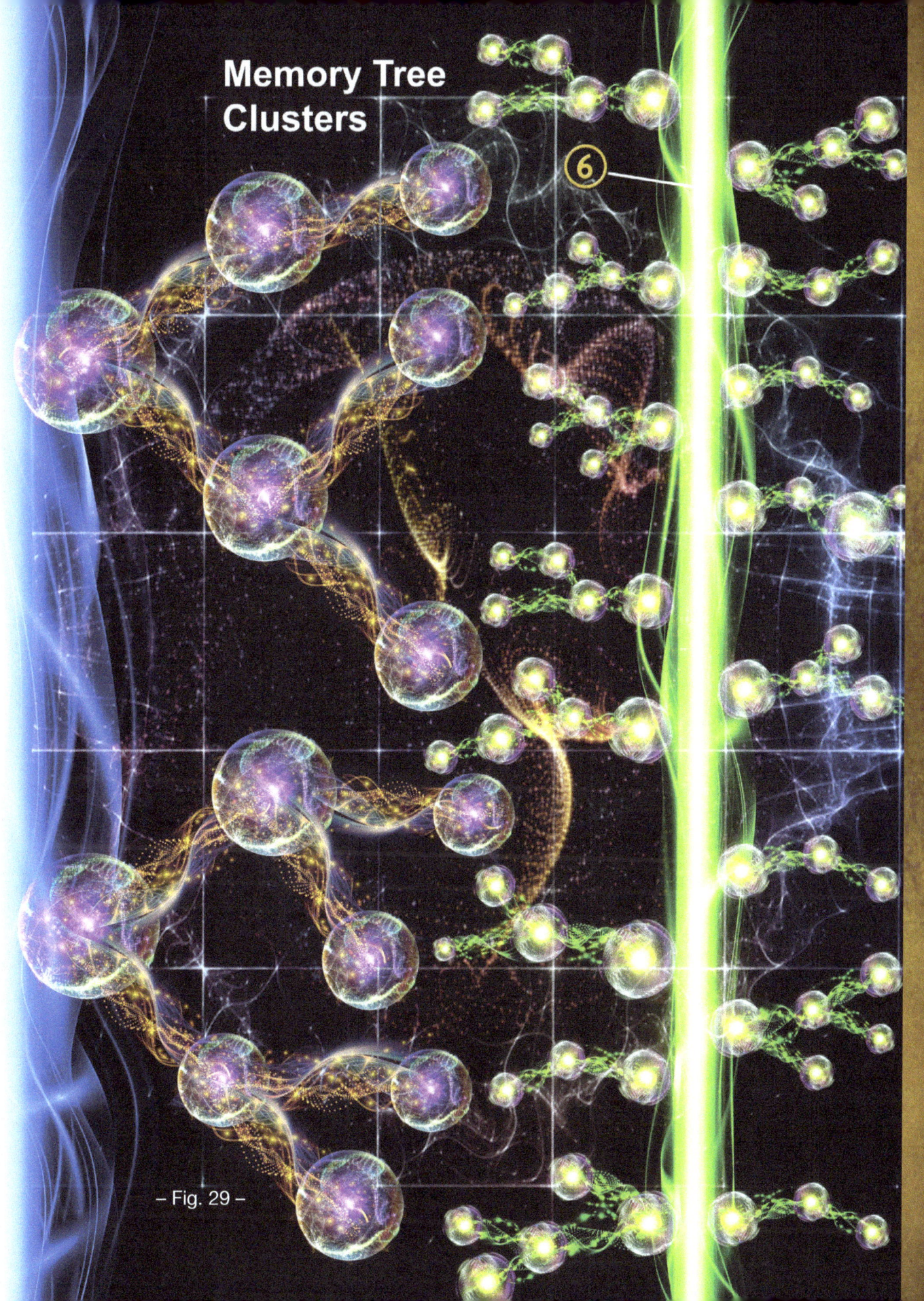

– Fig. 29 –

Cognitive Engineering

For simplicity, the drawing only shows the node particles in each tree. Imagine each node particle connects a memory record to the tree. The trees in the drawing also only have a few memories for simplicity, but actual trees can have dozens.

The central point of Fig. 29 is that memory trees with similar frequencies cluster around common memory centers.

As stated earlier, the records on memory trees are united by a common frequency. Fig. 21 depicts three kinds of memory trees whose unifying frequencies fall within three distinct frequency bands of experience. For illustration purposes, let's call these frequency bands visual, auditory and tactile.

Referring to Fig. 29, memory trees are shown to cluster around a blue visual center 1 in the foreground, and in the background, an orange auditory center 2 and green tactile center 6. These major energy centers are like mental superhighways which serve as conduits between the conscious being and its memory.

Root anchors 3 attach memory trees 4 and 5 to blue visual center 1. Orange and green centers 2 and 6 in the background are surrounded by clusters of memory trees which match their frequency, as shown by color.

Cognitive Engineering

 Root Anchors

Root anchors are the connection points between memory trees and 16 mental energy centers.

● Formation

Root anchors can be recent or past experiences, depending on the memory tree:

Forming New Trees	Growing Existing Trees	Reactivating Inert Memories
• When a new tree forms, the root anchor is a current experience which marks the beginning of the tree.	• New memories can attach themselves to existing active trees. In this case, the root anchor is the past experience which occurred when the tree was initially formed.	• New experiences can reactivate existing older inert memories. In this case, the root anchor is the current perception or stimulus which reactivates the inert memory.

● Persistence

Memory deactivation time is a function of charge. The higher the amplitude a memory tree has, the longer it will remain active and stick to the being.

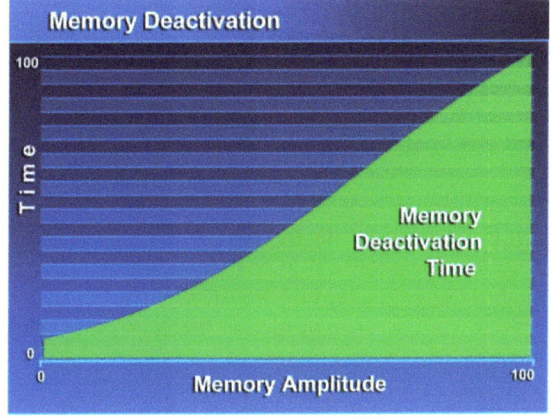

In contrast, low-amplitude trees can pass into and out of existence in a few seconds or minutes.

Since high-amplitude memories take longer to deactivate than low-amplitude memories, it is more probable that older memory trees will contain experiences originally recorded at higher amplitudes.

🟢 Deactivation

The root anchor governs the flow of applied awareness points from the conscious being into the tree. When the root anchor disappears, the memory tree loses its connection to the awareness energy that sustains it, and the tree implodes.

Memory trees can only persist when root anchors persist. When someone fully recalls a memory record, its particles and waves vanish. Recalling a memory tree's root anchor memory record causes the entire tree to unlatch itself from the being. The tree deactivates and the leaves fall off.

When a big memory tree implodes, one would expect to observe a noticeable drop in brainwave field voltage. As covered in the Neurodynamics chapter, a reduction in unconsciousness $U\downarrow$ will cause a decrease in brainwave voltage $V\downarrow$ assuming a constant neural resistance $R\leftrightarrow$. Per the Law of Conservation of Awareness, one would also expect to see a lift in conscious awareness $C\uparrow$, as U and C are reciprocals.

If someone partially discharges a root anchor, they may partially deactivate the memory tree by interfering with the flow of awareness points to the tree.

🎯 Summary

This reference chapter has described the 14 principles of macro memory behavior, including the recording, activation, deactivation, recall and reconsolidation of memory records, and the formation of memory trees. This information forms the backbone and essential building blocks for understanding the mechanics of the conscious being's unconscious mind.

Cognitive Engineering

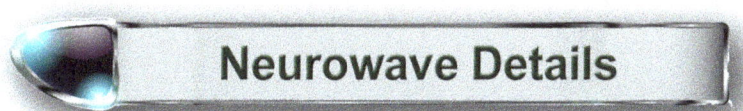
Neurowave Details

Different areas of the brain have many varieties of neuron types and axon lengths, creating a large array of different neurowave wavelengths. Neurowave characteristics are estimated below using typical average values for the electrophysiology of neurons.[18]

Neuron & Axon Characteristics		
Neuron	Pulse rise time (P)	1.2 ms (milliseconds)
	Peak pulse voltage	55 mV (millivolts)
	Average pulse current	5 nA (nanoamperes)
	Neuron cell diameter	25-50 micrometers
Axon	Electrochemical impulse transmission	50-60 MPH
	Axon length	50 micrometers to several feet

Axon Transmission Speed Examples (MPH)	50	55	60
Feet per microsecond for 1 MPH 0.00000146			
Axon speed feet per microsecond	0.0000733	0.0000807	0.0000880
Meters per foot 0.3937			
Axon signal speed micrometers per microsecond	28.87	31.76	34.65

Feet per mile = 5,280 • Microseconds per hour = 3,600,000,000

Axon Length Examples (micrometers)	100	200	400
Axon transmission time (microseconds)	3.464	6.298	11.545
Neuron Cell (soma)			
Soma diameter examples (microns)	15	25	40
Soma pulse rise time examples (microseconds)	38	50	70
Soma transmission time (microseconds)	38	50	70
Transmission Time (soma + axon)			
Microseconds	41	56	82
Seconds	.000041	.000056	.000082
Frequency (Hertz)	24,118	17,763	12,263
Frequency (K Hz)	24.12	17.76	12.26

---

[18] Source: *Electrophysiology of the Neuron*, Huguenard and McCormick, Oxford University Press

Present Time and Memory

As a being rises to higher states of awareness, it approaches present time more closely, moving out of the past (memory) and into the present (consciousness).

When a being becomes fully conscious, it is completely in present time, and has no active memory to filter its experience of reality and its own beingness.

Example: When a being reaches such a high state that they have temporarily forgotten how to use the body's sensory mechanisms, they find they have recovered enough of their awareness to perceive reality directly without the aid of the body's senses. The glass of iced tea you saw earlier with the body's eyes is no longer a glass. Instead, you find yourself marveling at the beautiful symmetry of the crystalline structure of its atomic latticework, and the sparkling patterns of energy that are holding it all together. In other words, you are in the glass.

In this state, almost all of your memories are inert. However, since the inert memories remain with you, they can always be reactivated. When this happens, you will notice yourself descending back into denser levels of unconsciousness.

Higher states of awareness can be made more sustainable by removing memory particles. A good place to start might be emotionally-charged, high-amplitude unpleasant memories, because their animation consumes the most awareness.

Contemporary memory reconsolidation theory offers some promising ideas along these avenues, which may represent a fruitful area for new research.

Cognitive Engineering

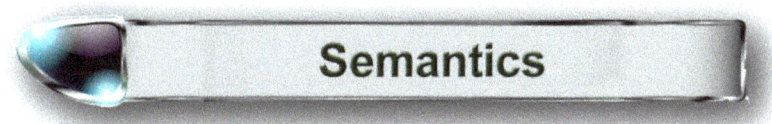

● Semantics

"Consciousness," "being" and "unconsciousness" are common words worn smooth by a million tongues, and their meanings are no longer sharp. A group of ten philosophers will have ten different interpretations for each one, and they can argue for days about what the words mean.

Our purpose here is not to debate the ultimate meaning of reality; it is to present a straightforward, practical system that produces specific, definite results for individuals.

Replacing these slippery words with a new nomenclature that limits them to precise meanings would be convenient, but new words would be cumbersome for readers deal with. Therefore, this text uses the common words, but their meanings are constrained as follows:

Being: A discrete unit of identity composed of awareness.

Conscious: Attentive, aware, cognizant, mindful, observing, perceiving, understanding. Technically, free awareness points.

Unconscious: Asleep, dormant, inert, unaware. Technically, applied awareness points.

Beings exist within a universal field of spirit, and interact with one another and the field. However, they are also sovereign individuals with their own unique memories, personalities, abilities, and behaviors. This distinction provides the foundation for a workable cognitive science for raising human awareness levels.

Conscious and unconscious: The Law of Conservation of Awareness is phrased using the common words *conscious* and *unconscious* to simplify understanding. The law says "the two states of awareness are conscious and unconscious." A more precise definition would be "the two states of awareness are free awareness points and applied awareness points."

Cognitive Engineering

● **Nomenclature**

For ease of comprehension, this book is written in standard English without special nomenclature. However, a shorthand nomenclature exists which researchers may wish to avail themselves of.

io	spirit / awareness;	i = free	o = applied	symbol = ϕ

pio awareness point– a point, particle, or packet of io (like a photon)

ion conscious being

bion embodied ion (e.g., human being)

iout external ion

ioff unconsciousness

bioff sleep

nion superconscious human being (*hyper sapiens*)

ioio awareness of awareness

hio higher awareness

ionic relating to awareness

igo continuous identity (multiple lifetimes)

iodyne life force

ionics cognitive physics

Cognitive Engineering

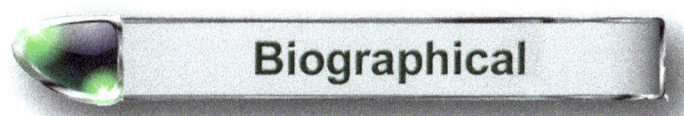

(30) Aurora borealis

Week of March 22, 1969

The widow of President John F. Kennedy, drops the name "Kennedy" after remarriage. She now uses Jacqueline Bouvier Onassis as her full name.

President Nixon presents a gold Commemorative Medal to Mrs. Walt Disney in recognition of the efforts of her late husband to make the world "a better place to live" 18 members of the Disney family were present in the White House state dining room and 200 children from Washington schools. They heard the President characterize Disney as a man who "never talked down to children."

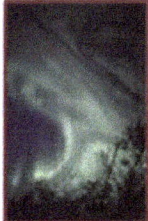

Sci-Fi - stranger than true! This week, the aurora borealis - northern lights, is seen over the eastern part of the country, with sightings reported as far south as Louisiana.

The Chicago Daily News reports that former heavyweight-boxing champion Muhammad Ali has been defrocked as a Black Muslim minister and thrown out of the sect. He was ousted by direct order of Elijah Muhammad, leader of the Black Muslims.

One out of every four American wives would pick another husband, according to a poll published in Family Circle. Only one in every 10 husbands, however, says he would have proposed to another girl if he had known as much about his wife then as he does now.

(31) OBE experiences
Catholic church – Manchester, Connecticut
Protestant chapel – Chapel of the Red Rocks, Sedona, Arizona
Hindu temple – Self Realization Fellowship, Phoenix, Arizona
Busy city – Los Angeles (night)
Placid forest – Sedona, Arizona (day)
Red rock canyons – Sedona, Arizona (day)
Mountain creeks – Sedona, Arizona (day)
Golf courses – Sedona, Arizona (night)
Barber shop – July 14, 1969
Basement – Rebirthing workshop, New Jersey
Bedrooms – Family home, summer 1969; Phoenix, Arizona, late 1980's (office)
Guest room – Bay Shore, July 13, 1969
Dorm rooms – Wesleyan University
Hotel rooms – The Pointe Resort, Phoenix, Arizona, 1989
Tents – Camping in Northern Arizona, mid-1980's

About the Author

J.L. Mee has had a rare opportunity to study the spirit / brain interface from an electrical engineering perspective for four decades. This unusual line of research has generated a new body of scientific work.

Mr. Mee is the founder of Cognitive Genetics Institute, where he directs its research and development programs.

www.ingramcontent.com/pod-product-compliance
Lightning Source LLC
Chambersburg PA
CBHW041219240426
43661CB00012B/1089